Challenges in Cybersecurity and Privacy – the European Research Landscape

RIVER PUBLISHERS SERIES IN SECURITY AND DIGITAL FORENSICS

Series Editors:

WILLIAM J. BUCHANAN
Edinburgh Napier University, UK

ANAND R. PRASAD
NEC, Japan

Indexing: All books published in this series are submitted to the Web of Science Book Citation Index (BkCI), to SCOPUS, to CrossRef and to Google Scholar for evaluation and indexing.

The "River Publishers Series in Security and Digital Forensics" is a series of comprehensive academic and professional books which focus on the theory and applications of Cyber Security, including Data Security, Mobile and Network Security, Cryptography and Digital Forensics. Topics in Prevention and Threat Management are also included in the scope of the book series, as are general business Standards in this domain.

Books published in the series include research monographs, edited volumes, handbooks and textbooks. The books provide professionals, researchers, educators, and advanced students in the field with an invaluable insight into the latest research and developments.

Topics covered in the series include, but are by no means restricted to the following:

- Cyber Security
- Digital Forensics
- Cryptography
- Blockchain
- IoT Security
- Network Security
- Mobile Security
- Data and App Security
- Threat Management
- Standardization
- Privacy
- Software Security
- Hardware Security

For a list of other books in this series, visit www.riverpublishers.com

Challenges in Cybersecurity and Privacy – the European Research Landscape

Editors

Jorge Bernal Bernabe
Antonio Skarmeta

University of Murcia, Spain

LONDON AND NEW YORK

Published 2019 by River Publishers
River Publishers
Alsbjergvej 10, 9260 Gistrup, Denmark
www.riverpublishers.com

Distributed exclusively by Routledge

4 Park Square, Milton Park, Abingdon, Oxon OX14 4RN
605 Third Avenue, New York, NY 10158

First published in paperback 2024

Challenges in Cybersecurity and Privacy – the European Research Landscape / by Jorge Bernal Bernabe, Antonio Skarmeta.

Routledge is an imprint of the Taylor & Francis Group, an informa business

Publisher's Note
The publisher has gone to great lengths to ensure the quality of this reprint but points out that some imperfections in the original copies may be apparent.

While every effort is made to provide dependable information, the publisher, authors, and editors cannot be held responsible for any errors or omissions.

ISBN: 978-87-7022-088-0 (hbk)
ISBN: 978-87-7004-357-1 (pbk)
ISBN: 978-1-003-33749-2 (ebk)

DOI: 10.1201/9781003337492

Contents

*Jorge Bernal Bernabe, Alejandro Molina, Antonio Skarmeta,
Stefano Bianchi, Enrico Cambiaso, Ivan Vaccari, Silvia Scaglione,
Maurizio Aiello, Rubén Trapero, Mathieu Bouet, Dallal Belabed,
Miloud Bagaa, Rami Addad, Tarik Taleb, Diego Rivera,
Alie El-Din Mady, Adrian Quesada Rodriguez, Cédric Crettaz,
Sébastien Ziegler, Eunah Kim, Matteo Filipponi, Bojana Bajic,
Dan Garcia-Carrillo and Rafael Marin-Perez*

7 CIPSEC-Enhancing Critical Infrastructure Protection with Innovative Security Framework **129**

Antonio Álvarez, Rubén Trapero, Denis Guilhot,
Ignasi García-Mila, Francisco Hernandez, Eva Marín-Tordera,
Jordi Forne, Xavi Masip-Bruin, Neeraj Suri, Markus Heinrich,
Stefan Katzenbeisser, Manos Athanatos, Sotiris Ioannidis,
Leonidas Kallipolitis, Ilias Spais, Apostolos Fournaris
and Konstantinos Lampropoulos

Danny S. Guamán, Manel Medina, Pablo López-Aguilar,
Hristina Veljanova, José M. del Álamo, Valentin Gibello,
Martin Griesbacher and Ali Anjomshoaa

Jorge Bernal Bernabe, Rafael Torres, David Martin,
Alberto Crespo, Antonio Skarmeta, Dave Fortune, Juliet Lodge,
Tiago Oliveira, Marlos Silva, Stuart Martin, Julian Valero
and Ignacio Alamillo

14 FutureTrust – Future Trust Services for Trustworthy Global Transactions   **285**

Detlef Hühnlein, Tilman Frosch, Jörg Schwenk,
Carl-Markus Piswanger, Marc Sel, Tina Hühnlein, Tobias Wich,
Daniel Nemmert, René Lottes, Stefan Baszanowski, Volker Zeuner,
Michael Rauh, Juraj Somorovsky, Vladislav Mladenov,
Cristina Condovici, Herbert Leitold, Sophie Stalla-Bourdillon,
Niko Tsakalakis, Jan Eichholz, Frank-Michael Kamm,
Jens Urmann, Andreas Kühne, Damian Wabisch, Roger Dean,
Jon Shamah, Mikheil Kapanadze, Nuno Ponte, Jose Martins,
Renato Portela, Çağatay Karabat, Snežana Stojičić,
Slobodan Nedeljkovic, Vincent Bouckaert, Alexandre Defays,
Bruce Anderson, Michael Jonas, Christina Hermanns,
Thomas Schubert, Dirk Wegener and Alexander Sazonov

15 LEPS – Leveraging eID in the Private Sector   303

Jose Crespo Martín, Nuria Ituarte Aranda,
Raquel Cortés Carreras, Aljosa Pasic, Juan Carlos Pérez Baún,
Katerina Ksystra, Nikos Triantafyllou, Harris Papadakis,
Elena Torroglosa and Jordi Ortiz

Preface

The global and hyper-connected society is experimenting an increasing number of Cybersecurity and Privacy issues. The widespread usage and development of ICT systems is expanding the number of attacks and leading to new kind of evolving cyber-threats, which ultimately undermines the possibilities of a trusted and dependable global digital society development.

Cyber-criminals are continuously shifting their cyber-attacks specially against cyber-physical systems and IoT, since they present additional vulnerabilities due to their constrained capabilities, their unattended nature and the usage of potential untrustworthiness components.

In this context, several cybersecurity and privacy challenges can be identified. Some of these challenges revolve around the autonomic cybersecurity management, orchestration and enforcement in heterogeneous and virtualized CPS/IoT and mobile ecosystems. Some other challenges are related to: cognitive detection and mitigation of evolving new kind of cyber-threats; the dynamic risk assessment and evaluation of cybersecurity, trustworthiness levels, privacy and legal compliance of ICT systems; the digital Forensics handling; the security intelligent and incident information exchange; cybersecurity and privacy tools and the associated usability and human factor. Similarly, regarding privacy and trust related challenges, four main global challenges can be identified, encompassing the reliable and privacy-preserving identity management, efficient and secure cryptographic mechanisms, Global trust management and privacy assessment.

Therefore, new holistic approaches, methodologies, techniques and tools are needed to cope with those issues, and mitigate cyberattacks, by employing novel cyber-situational awareness frameworks, risk analysis and modeling, threat intelligent systems, cyber-threat information sharing methods, advanced big-data analysis techniques as well as exploiting the benefits from latest technologies such as SDN/NFV and Cloud systems. In addition, novel privacy-preserving techniques, and crypto-privacy mechanisms, identity and eID management systems, trust services, and recommendations are needed to protect citizens' privacy while keeping usability levels.

The European Commission is facing the aforementioned cybersecurity and privacy challenges through different means, including the Horizon 2020 Research and Innovation program, and concretely, the European program H2020-EU.3.7, entitled *"Secure societies – Protecting freedom and security of Europe and its citizens",* and therefore, financing innovative projects that can cope with the increasing cyberthreat landscape.

This book presents and analyses 14 cybersecurity and privacy-related EU projects founded by that European program H2020-EU.3.7, encompassing: ANASTACIA, SAINT, FORTIKA, CYBECO, SISSDEN, CIPSEC, CS-AWARE. RED-Alert, Truessec.eu. ARIES, LIGHTest, CREDENTIAL, FutureTrust and LEPS.

The book is the result of a collaborative effort among relative ongoing European Research projects in the field of privacy and security as well as related cybersecurity fields, and it is intended to explain how these projects meet the main cybersecurity and privacy challenges faced in Europe. In the book we have invited to contribute with his knowledge some of the top cybersecurity and privacy experts and researcher from Europe.

The first introduction chapter identifies and describes 10 main cybersecurity and privacy research challenges presented and addressed in this book by 14 European research projects. In the book, each chapter is dedicated to a different funded European Research project and includes the project's overviews, objectives, and the particular research challenges that they are facing.

In addition, we have required each chapter' authors to provide, for his EU research project analysed, the research achievements on security and privacy, as well as the techniques, outcomes, and evaluations accomplished in the scope of the corresponding EU project.

The first part of the book, i.e. chapters from #2 to #10 describe 9 EU projects related to cybersecurity and how they face the challenges identified in Introduction section. Concretely: ANASTACIA, SAINT, FORTIKA, CYBECO, SISSDEN, CIPSEC, CS-AWARE. RED-Alert, Truessec.eu. The second part of the book, i.e. chapters from #11 to #15, describe 5 EU projects focused on privacy and Trust management. Namely, ARIES, LIGHTest, CREDENTIAL, FutureTrust and LEPS.

The idea of this book was originated after a successful clustering workshop entitled *"European projects Clustering workshop On Cybersecurity and Privacy (ECoSP 2018)"* collocated in ARES Conference – *13th International Conference on Availability, Reliability and Security*, held in Hamburg, Germany, where the EU projects analyzed in this book were presented and the attenders exchanged their views about the European research landscape on Security and privacy.

The chapters have been written for target both, researchers and engineers. Thus, after reading this book, academic researchers will have a proper understanding of current cybersecurity and privacy challenges to be solved in the coming years, and how they are being approached in different angles by several European research projects. Likewise, engineers will get to know the main enablers, technologies and tools that are being considered and implemented to deal with those main cybersecurity and privacy issues.

List of Contributors

Adamantios Koumpis, *University of Passau, Germany;*
E-mail: adamantios.koumpis@uni-passau.de

Adrian Quesada Rodriguez, *Mandat International, Research, Switzerland;*
E-mail: aquesada@mandint.org

Aitor Couce Vieira, *Institute of Mathematical Sciences (ICMAT), Spanish National Research Council (CSIC), Spain; E-mail: aitor.couce@icmat.es*

Alberto Crespo, *Atos Research and Innovation, Atos, Calle Albarracin 25, Madrid, Spain; E-mail: alberto.crespo@atos.net*

Alejandro Molina, *Department of Information and Communications Engineering, University of Murcia, Murcia, Spain;*
E-mail: alejandro.mzarca@um.es

Ales Černivec, *XLAB d.o.o., Slovenia; E-mail: ales.cernivec@xlab.si*

Alexander Sazonov, *National Certification Authority Rus CJSC (NCA Rus), 8A building 5, Aviamotornaya st., Moscow 111024, Russia;*
E-mail: sazonov@nucrf.ru

Alexandre Defays, *Arŋs Spikeseed, Rue Nicolas Bové 2B, 1253 Luxembourg, Luxembourg; E-mail: alexandre.defays@arhs-developments.com*

Alexandros Papanikolaou, *InnoSec, Greece;*
E-mail: a.papanikolaou@innosec.gr

Ali Anjomshoaa, *Digital Catapult, Research and Development, NW1 2RA, London, United Kingdom; E-mail: ali.anjomshoaa@ktn-uk.org*

Alie El-Din Mady, *United Technologies Research Center, Ireland;*
E-mail: madyaa@utrc.utc.com

Aljosa Pasic, *Atos Research and Innovation (ARI), Atos, Spain;*
E-mail: aljosa.pasic@atos.net

Anargyros Sideris, *Future Intelligence LTD, United Kingdom;*
E-mail: Sideris@f-in.co.uk

Anastasios Drosou, *Information Technologies Institute, Centre for Research & Technology Hellas, Greece; E-mail: drosou@iti.gr*

Andreas Kühne, *Trustable Limited, Great Hampton Street 69, Birmingham B18 6E, United Kingdom; E-mail: kuehne@trustable.de*

Angelo Consoli, *Eclexys Sagl, Via Dell Inglese 6, Riva San Vitale, Switzerland; E-mail: angelo.consoli@eclexys.com*

Antonio Álvarez, *ATOS SPAIN, Spain; E-mail: antonio.alvarez@atos.net*

Antonio Skarmeta, *Department of Information and Communications Engineering, University of Murcia, Murcia, Spain; E-mail: skarmeta@um.es*

Apostolos Fournaris, *University of Patras, Greece;*
E-mail: apofour@ece.upatras.gr

Arnolt Spyros, *InnoSec, Greece; E-mail: a.spyros@innosec.gr*

Bojana Bajic, *Archimede Solutions, Geneva, Switzerland;*
E-mail: bbajic@archimede.ch

Bruce Anderson, *Law Trusted Third Party Service (Pty) Ltd. (LAWTrust), 5 Bauhinia Street, Building C, Cambridge Office Park Veld Techno Park, Centurion 0157, South Africa; E-mail: bruce@LAWTrust.co.za*

Çağatay Karabat, *Turkiye Bilimsel Ve Tknolojik Arastirma Kurumu, Ataturk Bulvari 221, Ankara 06100, Turkey; E-mail: cagatay.karabat@tubitak.gov.tr*

Carl-Markus Piswanger, *Bundesrechenzentrum GmbH, Hintere Zollamtsstraße 4, A-1030 Vienna, Austria;*
E-mail: carl-markus.piswanger@brz.gv.at

Caroline Baylon, *AXA Technology Services, France;*
E-mail: caroline.baylon@axa.com

Cédric Crettaz, *Mandat International, Research, Switzerland;*
E-mail: ccrettaz@mandint.org

Chris Wills, *CARIS Research Ltd., United Kingdom;*
E-mail: ccwills@carisresearch.co.uk

Christina Hermanns, *Federal Office of Administration (Bundesverwaltungsamt), Barbarastr. 1, 50735 Cologne, Germany;*
E-mail: christina.hermanns@bva.bund.de

Cristi Potlog, *SIVECO Romania SA, Romania;*
E-mail: cristi.potlog@siveco.ro

Cristina Condovici, *Ruhr Universität Bochum, Universitätsstraße 150,*
44801 Bochum, Germany; E-mail: cristina.condovici@rub.de

Dallal Belabed, *THALES Communications & Security SAS, Gennevilliers,*
France; E-mail: dallal.belabed@thalesgroup.com

Damian Wabisch, *Trustable Limited, Great Hampton Street 69, Birmingham*
B18 6E, United Kingdom; E-mail: damian@trustable.de

Dan Garcia-Carrillo, *Department of Research & Innovation, Odin*
Solutions, Murcia, Spain; E-mail: dgarcia@odins.es

Daniel Abel, *Maven Seven Solutions Zrt., Hungary;*
E-mail: daniel.abel@maven7.com

Daniel Nemmert, *ecsec GmbH, Sudetenstraße 16, 96247 Michelau,*
Germany; E-mail: daniel.nemmert@ecsec.de

Danny S. Guamán, *1. Universidad Politécnica de Madrid, Departamento de*
Ingeniería de Sistemas Telemáticos, 28040, Madrid, Spain;
2. Escuela Politécnica Nacional, Departamento de Electrónica,
Telecomunicaciones y Redes de Información, 170525, Quito, Ecuador;
E-mail: ds.guaman@dit.upm.es

Dave Fortune, *Saher Ltd., United Kingdom; E-mail: dave@saher-uk.com*

David Martin, *GEMALTO, Czech Republic;*
E-mail: martin.david@gemalkto.com

David Ríos Insua, *Institute of Mathematical Sciences (ICMAT), Spanish*
National Research Council (CSIC), Spain; E-mail: david.rios@icmat.es

Dawn Branley-Bell, *Psychology, University of Northumbria at Newcastle,*
United Kingdom; E-mail: dawn.branley-bell@northumbria.ac.uk

Deepak Subramanian, *AXA Technology Services, France;*
E-mail: deepak.subramanian@axa.com

Denis Guilhot, *WORLDSENSING Limited, Spain;*
E-mail: dguilhot@worldsensing.com

Detlef Hühnlein, *ecsec GmbH, Sudetenstraße 16, 96247 Michelau,*
Germany; E-mail: detlef.huhnlein@ecsec.de

Diego Rivera, *R&D Department, Montimage, 75013, Paris, France;*
E-mail: diego.rivera@montimage.com

Dimitrios Tzovaras, *Information Technologies Institute, Centre for Research & Technology Hellas, Greece; E-mail: tzovaras@iti.gr*

Dirk Wegener, *German Federal Information Technology Centre (Informationstechnikzentrum Bund, ITZBund), Waterloostr. 4, 30169 Hannover, Germany; E-mail: dirk.dirkwegener@itzbund.de*

Edgardo Montes de Oca, *Montimage Eurl, 39 rue Bobillot, Paris, France;*
E-mail: edgardo.montesdeoca@montimage.com

Elena Torroglosa, *Department of Information and Communications Engineering, Faculty of Computer Science, University of Murcia, Murcia, Spain; E-mail: emtg@um.es*

Enrico Cambiaso, *National Research Council (CNR-IEIIT) – Via De Marini 6 – 16149 Genoa, Italy; E-mail: enrico.cambiaso@ieiit.cnr.it*

Eunah Kim, *Device Gateway SA, Research and Development, Switzerland;*
E-mail: eunah.kim@devicegateway.com

Eva Marín-Tordera, *Universitat Politècnica de Catalunya, Spain;*
E-mail: eva@ac.upc.edu

Evangelos K. Markakis, *Department of Informatics Engineering, Technological Educational Institute of Crete, Greece;*
E-mail: Markakis@pasiphae.teicrete.gr

Evangelos Pallis, *Department of Informatics Engineering, Technological Educational Institute of Crete, Greece; E-mail: Pallis@pasiphae.teicrete.gr*

Francisco Hernandez, *WORLDSENSING Limited, Spain;*
E-mail: fhernandez@worldsensing.com

Frank-Michael Kamm, *Giesecke & Devrient GmbH, Prinzregentestraße 159, 81677 Munich, Germany; E-mail: frank-michael.kamm@gi-de.com*

Georgios Sakellariou, *Department of Applied Informatics, University of Macedonia, Greece; E-mail: geosakel@uom.edu.gr*

Gerald Quirchmayr, *University of Vienna – Faculty of Computer Science, Austria; E-mail: gerald.quirchmayr@univie.ac.at*

Harris Papadakis, *University of the Aegean, i4m Lab (Information Management Lab) and Hellenic Mediterranean University, Greece;*
E-mail: adanar@atlantis-group.gr

Heiko Roßnagel, *Fraunhofer IAO, Fraunhofer Institute of Industrial Engineering IAO, Nobelstr. 12, 70569 Stuttgart, Germany; E-mail: heiko.roßnagel@iao.fraunhofer.de*

Herbert Leitold, *A-SIT, Seidlgasse 22/9, A-1030 Vienna, Austria; E-mail: herbert.leitold@a-sit.at*

Hristina Veljanova, *University of Graz, Institute of Philosophy and Institute of Sociology, 8010, Graz, Austria; E-mail: hristina.veljanova@uni-graz.at*

Ignacio Alamillo, *University of Murcia, Murcia, Spain; E-mail: ignacio.alamillod@um.es*

Ignasi García-Mila, *WORLDSENSING Limited, Spain; E-mail: igarciamila@worldsensing.com*

Ilias Spais, *AEGIS IT RESEARCH LTD, United Kingdom; E-mail: hspais@aegisresearch.eu*

Ioannis Mavridis, *Department of Applied Informatics, University of Macedonia, Greece; E-mail: mavridis@uom.edu.gr*

Ivan Vaccari, *National Research Council (CNR-IEIIT) – Via De Marini 6 – 16149 Genoa, Italy; E-mail: ivan.vaccari@ieiit.cnr.it*

Jan Eichholz, *Giesecke & Devrient GmbH, Prinzregentestraße 159, 81677 Munich, Germany; E-mail: jan.eichholz@gi-de.com*

Jart Armin, *CyberDefcon BV, Herengracht 282, 1016 BX Amsterdam, The Netherlands; E-mail: jart@cyberdefcon.com*

Jens Urmann, *Giesecke & Devrient GmbH, Prinzregentestraße 159, 81677 Munich, Germany; E-mail: jens.urmann@gi-de.com*

John M. A. Bothos, *National Center for Scientific Research "Demokritos", Patr. Gregoriou E. & 27 Neapoleos Str, Athens, Greece; E-mail: jbothos@iit.demokritos.gr*

John Sören Pettersson, *Karlstad University, Sweden; E-mail: john_soren.pettersson@kau.se*

Jon Shamah, *European Electronic Messaging Association AISBL, Rue Washington 40, Bruxelles 1050, Belgium; E-mail: jon.shamah@eema.org*

Jordi Forne, *Universitat Politècnica de Catalunya, Spain; E-mail: jforne@entel.upc.edu*

Jordi Ortiz, *Department of Information and Communications Engineering, Faculty of Computer Science, University of Murcia, Murcia, Spain; E-mail: jordi.ortiz@um.es*

Jörg Schwenk, *Ruhr Universität Bochum, Universitätsstraße 150, 44801 Bochum, Germany; E-mail: jorg.schwenk@rub.de*

Jorge Bernal Bernabe, *Department of Information and Communications Engineering, University of Murcia, Murcia, Spain; E-mail: jorgebernal@um.es*

Jose Crespo Martín *Atos Research and Innovation (ARI), Atos, Spain; E-mail: jose.crespomartin.external@atos.net*

José M. del Álamo, *Universidad Politécnica de Madrid, Departamento de Ingeniería de Sistemas Telemáticos, 28040, Madrid, Spain; E-mail: jm.delalamo@upm.es*

Jose Martins, *Multicert – Servicos de Certificacao Electronica SA, Lagoas Parque Edificio 3 Piso 3, Porto Salvo 2740 266, Portugal; E-mail: jose.martinsmulticert.com*

José Vila, *Devstat, Spain; E-mail: jvila@devstat.com*

Juan Carlos Pérez Baún, *Atos Research and Innovation (ARI), Atos, Spain; E-mail: juan.perezb@atos.net*

Juha Röning, *University of Oulu – Faculty of Information Technology and Electrical Engineering, Finland; E-mail: juha.röning@oulu.fi*

Julian Valero, *University of Murcia, Murcia, Spain; E-mail: julivale@um.es*

Juliano Efson Sales, *University of Passau, Germany; E-mail: juliano-sales@uni-passau.de*

Juliet Lodge, *Saher Ltd., United Kingdom; E-mail: juliet@saher-uk.com*

Juraj Somorovsky, *Ruhr Universität Bochum, Universitätsstraße 150, 44801 Bochum, Germany; E-mail: juraj.somorovsky@rub.de*

Katerina Ksystra, *University of the Aegean, i4m Lab (Information Management Lab), Greece; E-mail: katerinaksystra@gmail.com*

Katsiaryna Labunets, *Faculty of Technology, Policy and Management, Delft University of Technology, The Netherlands; E-mail: K.Labunets@tudelft.nl*

Kim Gammelgaard, *RheaSoft, Denmark; E-mail: kim@rheasoft.dk*

Konstantinos Lampropoulos, *University of Patras, Greece; E-mail: klamprop@ece.upatras.gr*

Konstantinos M. Giannoutakis, *Information Technologies Institute, Centre for Research & Technology Hellas, Greece; E-mail: kgiannou@iti.gr*

Konstantinos Rantos, *Eastern Macedonia and Thrace Institute of Technology, Department of Computer and Informatics Engineering, Greece; E-mail: krantos@teiemt.gr*

Laurentiu Vasiliu, *Peracton, Ireland; E-mail: laurentiu.vasiliu@peracton.com*

Leonidas Kallipolitis, *AEGIS IT RESEARCH LTD, United Kingdom; E-mail: lkallipo@aegisresearch.eu*

Manel Medina, *1. Universitat Politécnica de Catalunya, esCERT-inLab, 08034, Barcelona, Spain; 2. APWG European Union Foundation, Research and Development, 08012, Barcelona, Spain; E-mail: medina@ac.upc.edu*

Manos Athanatos, *Foundation for Research and Technology – Hellas, Greece; E-mail: athanat@ics.forth.gr*

Marc Sel, *PwC Enterprise Advisory, Woluwedal 18, Sint Stevens Woluwe 1932, Belgium; E-mail: marc.sel@be.pwc.com*

Markus Heinrich, *Technische Universität Darmstadt, Germany; E-mail: heinrich@seceng.informatik.tu-darmstadt.de*

Marlos Silva, *SONAE, Portugal; E-mail: mhsilva@sonae.pt*

Martin Griesbacher, *University of Graz, Institute of Philosophy and Institute of Sociology, 8010, Graz, Austria; E-mail: m.griesbacher@uni-graz.at*

Mathieu Bouet, *THALES Communications & Security SAS, Gennevilliers, France; E-mail: mathieu.bouet@thalesgroup.com*

Matteo Bregonzio, *3rd Place, Italy; E-mail: matteo.bregonzio@3rdplace.com*

Matteo Filipponi, *Device Gateway SA, Research and Development, Switzerland; E-mail: mfilipponi@devicegateway.com*

Maurizio Aiello, *National Research Council (CNR-IEIIT) – Via De Marini 6 – 16149 Genoa, Italy; E-mail: maurizio.mongelli@ieiit.cnr.it*

Michael Jonas, *Federal Office of Administration (Bundesverwaltungsamt), Barbarastr. 1, 50735 Cologne, Germany; E-mail: michael.jonas@bva.bund.de*

Michael Rauh, *ecsec GmbH, Sudetenstraße 16, 96247 Michelau, Germany; E-mail: michael.rauh@ecsec.de*

Mikheil Kapanadze, *Public Service Development Agency, Tsereteli Avenue 67A, Tbilisi 0154, Georgia; E-mail: mkapanadze@sda.gov.ge*

Miloud Bagaa, *Department of Communications and Networking, School of Electrical Engineering, Aalto University, Finland; E-mail: miloud.bagaa@aalto.fi*

Monica Florea, *SIVECO Romania SA, Romania; E-mail: Monica.Florea@siveco.ro*

Neeraj Suri, *Technische Universität Darmstadt, Germany; E-mail: suri@cs.tu-darmstadt.de*

Niko Tsakalakis, *University of Southampton, Highfield, Southampton S017 1BJ, United Kingdom; E-mail: niko.tsakalakis@soton.ac.uk*

Nikolaos Tsinganos, *Department of Applied Informatics, University of Macedonia, Greece; E-mail: tsinik@uom.edu.gr*

Nikolaos Zotos, *Future Intelligence LTD, United Kingdom; E-mail: Zotos@f-in.co.uk*

Nikos Triantafyllou, *University of the Aegean, i4m Lab (Information Management Lab), Greece; E-mail: triantafyllou.ni@gmail.com*

Nikos Vassileiadis, *Trek Consulting, Greece; E-mail: n.vasileiadis@trek-development.eu*

Nuno Ponte, *Multicert – Servicos de Certificacao Electronica SA, Lagoas Parque Edificio 3 Piso 3, Porto Salvo 2740 266, Portugal; E-mail: nuno.pontemulticert.com*

Nuria Ituarte Aranda *Atos Research and Innovation (ARI), Atos, Spain; E-mail: nuria.ituarte@atos.net*

Oscar Garcia, *Information Catalyst, Spain; E-mail: oscar.garcia@informationcatalyst.com*

Pablo López-Aguilar, *APWG European Union Foundation, Research and Development, 08012, Barcelona, Spain; E-mail: pablo.lopezaguilar@apwg.eu*

Pamela Briggs, *Psychology, University of Northumbria at Newcastle, United Kingdom; E-mail: p.briggs@northumbria.ac.uk*

Panayotis Fouliras, *Department of Applied Informatics, University of Macedonia, Greece; E-mail: pfoul@uom.edu.gr*

Peter Hamm, *Goethe University Frankfurt, Germany; E-mail: peter.hamm@m-chair.de*

Peter Pollner, *MTA-ELTE Statistical and Biological Physics Research Group, Hungary; pollner@angel.elte.hu*

Rafael Marin-Perez, *Department of Research & Innovation, Odin Solutions, Murcia, Spain; E-mail: rmarin@odins.es*

Rafael Torres, *University of Murcia, Murcia, Spain; E-mail: rtorres@um.es*

Rami Addad, *Department of Communications and Networking, School of Electrical Engineering, Aalto University, Finland; E-mail: rami.addad@aalto.fi*

Raquel Cortés Carreras, *Atos Research and Innovation (ARI), Atos, Spain; E-mail: raquel.cortes@atos.net*

Renato Portela, *Multicert – Servicos de Certificacao Electronica SA, Lagoas Parque Edificio 3 Piso 3, Porto Salvo 2740 266, Portugal; E-mail: renato.portelamulticert.com*

René Lottes, *ecsec GmbH,Sudetenstraße 16, 96247 Michelau, Germany; E-mail: rene.lottes@ecsec.de*

Roger Dean, *European Electronic Messaging Association AISBL, Rue Washington 40, Bruxelles 1050, Belgium; E-mail: r.dean@eema.org*

Rubén Trapero, *Atos Research and Innovation, Atos, Calle Albarracin 25, Madrid, Spain; E-mail: ruben.trapero@atos.net*

Sébastien Ziegler, *Mandat International, Research, Switzerland; E-mail: sziegler@mandint.org*

Shmuel Bar, *IntuView, Israel; E-mail: sbar@intuview.com*

Silvia Scaglione, *National Research Council (CNR-IEIIT) – Via De Marini 6 – 16149 Genoa, Italy; E-mail: silvia.scaglione@ieiit.cnr.it*

Slobodan Nedeljkovic, *Ministarstvo unutrašnjih poslova Republike Srbije, Kneza Miloša 103, Belgrade 11000, Serbia; E-mail: slobodan.nedeljkovic@mup.gov.rs*

Snežana Stojičić, *Ministarstvo unutrašnjih poslova Republike Srbije, Kneza Miloša 103, Belgrade 11000, Serbia; E-mail: snezana.stojicic@mup.gov.rs*

Sofia Tsekeridou, *Intrasoft International, Greece; E-mail: Sofia.Tsekeridou@intrasoft-intl.com*

Sophie Stalla-Bourdillon, *University of Southampton, Highfield, Southampton S017 1BJ, United Kingdom; E-mail: sophie.stalla-bourdillon@soton.ac.uk*

Sotiris Ioannidis, *Foundation for Research and Technology – Hellas, Greece; E-mail: sotiris@ics.forth.gr*

Stavros Salonikias, *Department of Applied Informatics, University of Macedonia, Greece; E-mail: salonikias@uom.edu.gr*

Stefan Baszanowski, *ecsec GmbH, Sudetenstraße 16, 96247 Michelau, Germany; E-mail: stefan.baszanowski@ecsec.de*

Stefan Katzenbeisser, *Technische Universität Darmstadt, Germany; E-mail: katzenbeisser@seceng.informatik.tu-darmstadt.de*

Stefan Schiffner, *University of Luxemburg; E-mail: Stefan.schiffner@uni.lu*

Stefano Bianchi, *Research & Innovation Department, SOFTECO SISMAT SRL, Di Francia 1 – WTC Tower, 16149, Genoa, Italy; E-mail: stefano.bianchi@softeco.it*

Stephan Krenn, *AIT Austrian Institute of Technology GmbH, Austria; E-mail: stephan.krenn@ait.ac.at*

Stuart Martin, *Office of the Police and Crime Commissioner for West Yorkshire, (POOC), West Yorkshire, United Kingdom; E-mail: stuart.martin@westyorkshire.pnn.police.uk*

Sven Wagner, *University Stuttgart, Institute of Human Factors and Technology Management, Allmandring 35, 70569 Stuttgart, Germany; E-mail: sven.wagner@iat.uni-stuttgart.de*

Syed Naqvi, *Birmingham City University, United Kingdom; E-mail: Syed.Naqvi@bcu.ac.uk*

Tarik Taleb, *Department of Communications and Networking, School of Electrical Engineering, Aalto University, Finland; E-mail: tarik.taleb@aalto.fi*

Thomas Schaberreiter, *University of Vienna – Faculty of Computer Science, Austria; E-mail: thomas.schaberreiter@univie.ac.at*

Thomas Schubert, *Federal Office of Administration (Bundesverwaltungsamt), Barbarastr. 1, 50735 Cologne, Germany; E-mail: thomas.schubert@bva.bund.de*

Tiago Oliveira, *SONAE, Portugal; E-mail: tioliveira@sonae.pt*

Tilman Frosch, *Ruhr Universität Bochum, Universitätsstraße 150, 44801 Bochum, Germany; E-mail: tilman.frosch@rub.de*

Tina Hühnlein, *ecsec GmbH, Sudetenstraße 16, 96247 Michelau, Germany; E-mail: tina.huhnlein@ecsec.de*

Tobias Wich, *ecsec GmbH, Sudetenstraße 16, 96247 Michelau, Germany; E-mail: tobias.wich@ecsec.de*

Valentin Gibello, *University of Lille, CERAPS – Faculty of Law, 59000, Lille, France; E-mail: valentin.gibello@univ-lille.fr*

Vassilis Chatzigiannakis, *Intrasoft International, Greece; E-mail: Vassilis.Chatzigiannakis@intrasoft-intl.com*

Veronika Kupfersberger, *University of Vienna – Faculty of Computer Science, Austria; E-mail: veronika.kupfersberger@univie.ac.at*

Vincent Bouckaert, *Arηs Spikeseed, Rue Nicolas Bové 2B, 1253 Luxembourg, Luxembourg; E-mail: vincent.bouckaert@arhs-developments.com*

Vladislav Mladenov, *Ruhr Universität Bochum, Universitätsstraße 150, 44801 Bochum, Germany; E-mail: vladislav.mladenov@rub.de*

Volker Zeuner, *ecsec GmbH, Sudetenstraße 16, 96247 Michelau, Germany; E-mail: volker.zeuner@ecsec.de*

Waqar Asif, City, *University of London, United Kingdom; E-mail: Waqar.Asif@city.ac.uk*

Wolter Pieters, *Faculty of Technology, Policy and Management, Delft University of Technology, The Netherlands; E-mail: W.Pieters@tudelft.nl*

Xavi Masip-Bruin, *Universitat Politècnica de Catalunya, Spain; E-mail: xmasip@ac.upc.edu*

Yannis Nikoloudakis, *Department of Informatics Engineering, Technological Educational Institute of Crete, Greece; E-mail: Nikoloudakis@pasiphae.teicrete.gr*

Yolanda Gómez, *Devstat, Spain; E-mail: ygomez@devstat.com*

List of Figures

List of Tables

List of Abbreviations

AAA	Authentication, Authorization and Accounting
ABAC	Attribute Based Access Control
ABC	Anti-Bot Code of Conduct
ABC	Attribute Based Credentials
ACS	Anonymous Credential Systems
AD	Architectural Description
AI	Artificial Intelligence
API	Application Programming Interface
AQDRS	Regional System of Detection of Air Quality
ASN	Anonymous Communication Networks
ASR	Automatic Speech Recognition
ATHEX	Athens Exchange Group
BAID	Bi-Agent Influence Diagram
C&C	Command & Control
CEF	Connecting Europe Facility
CEP	Complex Event Processing
CERT	Community Emergency Response Team
CERTs	Community Emergency Response Teams
CI, CIs	Critical Infrastructure, Critical Infrastructures
CIDR	Classless Inter-Domain Routing
CIPIs	Critical Infrastructure Performance Indicator
CMS	Compliance Management Service
CNN	Convolutional Neural Network
Conpo	Honeypot, http://conpot.org/
CORAS	Name of the risk analysis method developed by Lund, Solhaug and Stølen
CPS	Cyber Physical Systems
CSIRTs	Computer Security Incident Response Teams
CTI	Cyber Threat Intelligence
CYBECO	Accronym of the H2020 project "Supporting Cyberinsurance from a Behavioural Choice Perspective"

DDoS	Distributed Denial of Service
Dionaea	Honeypot, https://github.com/DinoTools/dionaea
DKMS	Decentralized Key Management System
DNP3	Distributed Network Protocol
DoS	Denial of Service
DoS	Denial of Service Attacks
DoW	Description of Work
DSI	Digital Service Infrastructure
DSM	Digital Single Market
DSS	Decision Support System
EBA	Eisenbahnbundesamt (Railway Federal Office)
EC	European Commission
EC3	European Cybercrime Centre
EHEA	European Higher Education Area
EIA	Ethical Impact Assessment
eID	electronic Identity
eID	electronic identification
eIDAS	electronic IDentification, Authentication and trust Services
ELTA	Hellenic Post
ENISA	European Union Agency for Network and Information Security
EU	European Union
FICORA	Finnish communications regulatory authority
FP7	7th Framework Programme
FPGA	Field Programmable Gate Array
G20	Group of 20
GDP	Gross Domestic Product
GDPR	General Data Protection Regulation (EU)
GSM	SAINT's Global Security Map
GW	Gateway
H2020	Horizon 2020
HIDS	Host Intrusion Detection Systems
HMI	Human-Machine Interface
HTTP	Hypertext Transfer Protocol
ICS	Integrated Computer Solution
ICT	Information and Communication Technology
ICTs	Information Communication Technologies
IdM	Identity Management
IdP	Identity Provider

IDPS	Intrusion Detection and Prevention System
IDS	Intrusion Detection System
IMG	Industry Monitoring Group
IMPACT	International Multilateral Partnership Against Cyber Threats
IoT	Internet of Things
IP	Internet Protocol
IP, IPs	Internet Protocol, IP addresses
IPS	Intrusion Prevention System
ISAC	Information Sharing and Analysis Center
ISF	Information Security Forum
ISO	International Organization for Standardization
ISPs	Internet Service Providers
ISS	Interconnection Supporting Service
IT	Information Technology
ITU	International Telecommunication Union
ITU-GCA	IUT Global Cyber-security Agenda
JBPM	Java Business Process Model
JSON	JavaScript Object Notation
JWT	JASON Web Token
Kippo	Honeypot, https://github.com/desaster/kippo
KPI	Key Performance Indicator
KPIs	Key Performance Indicators
LEA	Law Enforcement Agency
LEAs	Law Enforcement Agencies
LEPS	Leveraging eID in the Private Sector
LoA	Level of Assurance
MDM	Data Management systems
ML	Meta-Learning
MMT	Montimage Monitoring Tool
Modbus	Communications protocol, http://www.modbus.org/
MS	Member State
MSPL	Medium Security Policy Language
NFC	Near Field Communication
NFV	Network Function Virtualization
NHS	Britain's National Health Service
NIDS	Network Intrusion Detection Systems
NIS	Directive EU Directive on Security of Network and Information Systems
NLP	Natural Language Processing

NTA	Network Traffic Analysis
OC	Operations Centre
OPC	UA Open Platform Communications
OPC	Open Platform Communications
OSGi	Open Services Gateway initiative
OT	Open Platform Communications Unified Architecture
PAP	Policy Administration Point
PC	Personal Computer
PCAP	Packet Capture
PDP	Policy Decision Point
PEP	Policy Enforcement Point
PHR	Patient Healthcare Record
PIP	Policy Information Point
PLCs	Programmable Logic Controller
RDBMS	Relational Database Management System
ROI	Return of Investment
RPN	Region Proposal Network
RTNTA	Real Time Network Traffic Analyzer
SAINT	Systemic Analyser In Network Threats
SAML	Security Assertion Mark-up Language
SCADA	Supervisory Control and Data Acquisition
SDA	Slow DoS Attack
SDN	Software Defined Networks
SEARS	Social Engineering Attack Recognition System
SFC	Service Function Chain
SIEM	Security Information and Event Management
SISSDEN	Secure Information Sharing Sensor Delivery event Network
SLA	Service-Level Agreement
SMA	Semantic Multimedia Analysis
SME	Small and Medium Enterprises
SNA	Social Network Analysis
SoC	System on-Chip
SP	Service Provider
SQL	Structured Query Language
SSH	Secure Shel
SSI	Self-Sovereign Identity
STIX/TAXII	Structured Threat Information eXpression/rusted Automated Exchange of Indicator Information
UI	User Interface

UK	United Kingdom
URL	Uniform Resource Locator
URLs	Uniform Resource Locators
US	United States of America
USB	Universal Serial Bus
UTM	Unified Thread Management
VHDL	VHSIC Hardware Description Language
VHSIC	Very High Speed Integrated Circuit
VLAN	Virtual Local Area Network
VM	Virtual Machine
VNF	Virtual Network Functions
VPS	Virtual Private Server
VSA	Virtual Security Appliance
WPAN	Wireless Personal AreaNetworks
WSDL	Web Services Description Language
XL-SIEM	Cross-Layer Security Information and Event Management
ZKP	Zero Knowledge Proof

1

Introducing the Challenges in Cybersecurity and Privacy: The European Research Landscape

Jorge Bernal Bernabe and Antonio Skarmeta

Department of Information and Communications Engineering,
University of Murcia, Murcia, Spain
E-mail: jorgebernal@um.es; skarmeta@um.es

The continuous, rapid and widespread usage of ICT systems, the constrained and large-scale nature of certain related networks such as IoT (Internet of Things), the autonomous nature of upcoming systems, as well as the new cyber-threats appearing from new disruptive technologies, are given rise to new kind of cyberattacks and security issues. In this sense, this book chapter categorises and presents 10 current main cybersecurity and privacy research challenges, as well as 14 European research projects in the scope of cybersecurity and privacy, analysed further throughout this book, that are addressing these challenges.

1.1 Introduction

The widespread usage and development of ICT systems is leading to new kind of cyber-threats. Cyberattacks are continuously emerging and evolving, exploiting disruptive systems and technologies such as Cyber Physical Systems (CPS)/IoT, virtual technologies, clouds, mobile systems/networks, autonomous systems (e.g. drones, vehicles). Cyber attackers are continuously improving their techniques to come up with stealth and sophisticated attacks, especially against IoT, since these environments suffer additional vulnerabilities due to their constrained capabilities, their unattended nature and the usage of potential untrustworthiness components. Similarly, identity-theft, fraud, personal data leakages, and other related cyber-crimes are continuously evolving, causing important damages and privacy problems for European citizens in both virtual and physical scenarios.

In this evolving cyber-threat landscape, we have identified 10 main cybersecurity and privacy research challenges (described in Section 2 of this chapter):

1. Interoperable and scalable security management in heterogeneous ecosystems
2. Autonomic security orchestration and enforcement in softwarized and virtualized IoT/CPS systems and mobile environments
3. Cognitive detection and mitigation of evolving new kind of cyber-threats
4. Dynamic Risk assessment and evaluation of cybersecurity, trustworthiness levels, privacy and legal compliance of ICT systems
5. Digital Forensics handling, security intelligent and incident information exchange
6. Cybersecurity and privacy tools for end-users and SMEs. The usability and human factor challenges
7. Reliable and privacy-preserving physical and virtual identity management
8. Efficient and secure cryptographic mechanisms to strengthen confidentiality and privacy
9. Global trust management of eID and related services
10. Privacy assessment, run-time evaluation of the quality of security and privacy risks

To meet those challenges, new holistic approaches, methodologies, techniques and tools are needed to prevent and mitigate cyberattacks by employing novel cyber-situational awareness frameworks, risk analysis and modelling tools, threat intelligent systems, cyber-threat information sharing methods, advanced big-data analysis techniques as well as new solutions that can exploit the benefits brought from latest technologies such as SDN/NFV and Cloud systems. In addition, novel privacy-preserving techniques, and crypto-privacy mechanisms, identity and eID management systems, trust services, and recommendations are needed to protect citizens' privacy while keeping usability levels.

The European Commission is addressing the aforementioned challenges through different means, including the Horizon 2020 Research and

Innovation program, thereby financing innovative research projects that can cope with the increasing cyberthreat landscape.

In this sense, the cybersecurity strategy of the European Union is summarized in 5 strategic priorities "An Open, Safe and Secure Cyberspace" [1]

– *Achieving Cyber resilience;*
– *Reducing cybercrime;*
– *Developing a cyber defense policy and capabilities related to the Common Security and Defense Policy (CSDP);*
– *Developing the industrial and technological resources for cybersecurity;*
– *Establishing a coherent international cyberspace policy for the European Union that promoted core EU values.*

Namely, the European program H2020-EU.3.7 [2] – "Secure societies – Protecting freedom and security of Europe and its citizens", budget with 1694.60 million, is addressing those cybersecurity and privacy challenges. The general objective in that program is "*to foster secure European societies in a context of unprecedented transformations and growing global interdependencies and threats, while strengthening the European culture of freedom and justice.*"

Thus, the H2020-EU.3.7 program is addressing the global challenge about "*undertaking the research and innovation activities needed to protect our citizens, society and economy as well as our infrastructures and services, our prosperity, political stability and wellbeing.*" Namely, this programme [3] aims:

- "*to enhance the resilience of our society against natural and man-made disasters, ranging from the development of new crisis management tools to communication interoperability, and to develop novel solutions for the protection of critical infrastructure;*
- *to fight crime and terrorism ranging from new forensic tools to protection against explosives;*
- *to improve border security, ranging from improved maritime border protection to supply chain security and to support the Union's external security policies including through conflict prevention and peace building;*
- *and to provide enhanced cybersecurity, ranging from secure information sharing to new assurance models.*"

In this context, this book presents and analyses 14 cybersecurity and privacy-related EU projects founded by this H2020 program, encompassing: ANASTACIA, SAINT, FORTIKA, CYBECO, SISSDEN, CIPSEC, CS-AWARE. RED-Alert, Truessec.eu. ARIES, LIGHTest, CREDENTIAL, FutureTrust. For further information about other H2020 EU projects funded under this H2020-EU.3.7 the reader is refereed to [2].

Each chapter in the book is dedicated to a different funded European Research project and includes the project's overviews, objectives, and the particular research challenges, among the ones identified above, that they are facing. In addition, each EU research project in his corresponding chapter describes its research achievements on security and privacy, as well as the techniques, outcomes, and evaluations accomplished in the scope of the corresponding EU project.

The idea of this book was originated after a successful clustering workshop entitled *"European projects Clustering workshop On Cybersecurity and Privacy (ECoSP 2018)"* [4] collocated in ARES Conference – 13th International Conference on Availability, Reliability and Security, where the EU projects analyzed in this book were presented and the attenders exchanged their views about the European research landscape on Security and privacy.

The rest of this chapter is structured as follows. Section 2 presents the main security and privacy research challenges. Section 3 is devoted to the introduction of the main H2020 EU projects covered in this book, and the main challenges, among the ones identified in Section 2, that each project is facing. Section 4 concludes this chapter.

1.2 Cybersecurity and Privacy Research Challenges

The Ponemon Institute in a recent study [23], identified the Cyber threats with the greatest risk: Cyber warfare or cyber terrorism, Breaches involving high-value information, Nation-state attackers, Breaches that damage critical infrastructure, Breaches that disrupt business and IT processes, Emergence of cyber syndicates, Stealth and sophistication of cyber attackers, Emergence of hacktivism, Breaches involving large volumes of data, Malicious or criminal insiders, Negligent or incompetent employees. The study highlights that Cyber warfare and cyber terrorism and breaches involving high-value information will have the greatest impact on organizations over the next three years.

These cyber-threats are especially notorious and dangerous when affecting IoT and CPS, where massive heterogenous, and potentially constrained, things are being added to the network, meaning additional potential vulnerabilities. In this regard, Roman et al. [19] identified the main "Challenges of Security & Privacy in Distributed Internet of Things". Namely, they provided and analysis of attacker models and threats and identified 7 main challenges in the design and deployment of the security mechanisms, including: Identity and Authentication, Access control, Protocol and Network Security, Privacy, Trust management, Governance, Fault tolerance.

Additionally, recently [22] identified the security and privacy threats in IoT at different network layers, including the major security vulnerabilities. In that paper authors highlighted the main aspects of the IoT ecosystem, such as, having legacy systems running in these platforms, the large number of devices, dynamicity, constrained nature, which are provoking new kind of threats. Likewise, [25] reviewed the IoT cybersecurity research, highlighting the data handling issues, standardization aspects, and research trends when IoT meets Cloud Computing and 5G technologies. Other research trends (Fault Tolerance Mechanism, Self-Management, IoT Forensics, Blokchain Embedded Cybersecurity Design) are also studied.

Besides, Backes et al. [24] identified their 8 most important challenges in IT security research. Including, (1) Security for Autonomous Systems, (2) Security in Spite of Untrustworthy Components, (3) Security Commensurate with Risk, (4) Privacy for Big Data, (5) Economic Aspects of IT Security, (6) Behaviour-related and Human Aspects of IT Security (7) Security of Cryptographic Systems against Powerful Attacks, (8) Detection and Reaction.

The characterization presented herein includes most of those security research challenges but, unlike their work, we use another perspective and for us some of their research challenges (such as economic aspects) are out of our main challenges, as they are not such important in our classification.

The main cybersecurity and privacy research challenges identified are described below. It should be noted that order of challenges does not have any relation with the order of importance or impact of the challenges.

1.2.1 Main Cybersecurity Research Challenges

1. **Interoperable and scalable security management in heterogeneous ecosystems**
   Security Management in fragmented and heterogeneous domains is still nowadays an open research challenge. This issue is exacerbated in CPS/IoT deployments which are comprised of heterogenous disparate

kind of devices and networks protocols/systems. Security management requires a holistic approach to deal with new types of wireless network technologies (e.g. 5G), potentially constrained networks (e.g. LPWANs), protocols and systems, that need to face the management of large and scalable deployments in any segment of network: RAN, Edge, Fog or Core segments.

The definition of security management policies to deal with heterogeneity and interoperability across domains, systems and networks, introduces several challenges related to the employed security models, the language and the level of abstraction required to govern the systems. In this regard, interoperability and contextual aspects in policies, particularities of managed systems domains, policy conflicts and resolution as well as dependencies in policies, are open research challenges that need to be solved. The policies should encompass not only security/privacy policies, but also QoS/SLA policies, network management policies (e.g. slicing, traffic filtering), operational and orchestration policies.

2. **Autonomic security orchestration and enforcement in softwarized and virtualized IoT/CPS systems and mobile networks**

   o *Holistic security orchestration*: New autonomic and context-awareness security orchestrators are needed, which can choregraph and enforce quickly and dynamically the proper defence mechanism (proactively or as countermeasure), according to the circumstances, in SDN/NFV-enabled systems. The orchestration will need to face the challenge to interface with diverse, heterogeneous and distributed IoT controllers, NFV-MANO (Management and Orchestration) orchestrators, Fog-Edge entities, SDN controllers, thereby enforcing dynamically the security enablers in the network/systems.

   o *Virtualized and Softwarized security management*: current defences of network operators and companies are mainly based on hardware appliances. Naturally, the hardware appliances have fixed location that must be chosen by the ISP smartly. These hardware appliances can be deployed on-premises or outsourced, and the packets/flows are redirected to these hardware appliances. Using the virtualization enabled by SDN and NFV allows a quick instantiation of VMs in the adequate location. Indeed, the lack of elasticity can be easily handled by Security Virtual Network Function (VNF) functions that can be chained and placed on-demand according to the incoming attacks.

However, it is challenging to manage the orchestration and placement of multiple VNFs on an NFV Infrastructure at large scale, either at the core of at the edge of the network, while dealing with scalability and security issues and additional threats that raise from the fact of using a virtualized environment.

○ *Selection of the adequate mitigation plan*: and fast enforcement of the defined policies are challenging processes that require a lot of efforts and time. The orchestration and the enforcement of the adequate countermeasures in a short time, and without affecting the Quality of Service (QoS), introduce several challenges that must be duly considered. Also, the definition and enforcement of mitigation plans while reducing the deployment cost and by taking into account the limitations in existing infrastructure clouds, the system/network status and are open research questions that needs to be addressed.

○ *Lightweight Security enablers and protocols for IoT/CPS systems*: Traditional security enablers and protocols, encompassing Authentication, Authorization and Accounting (AAA), Channel protection protocols, network filtering, deep packet inspection, intrusion detection..., need to be evolved and adapted to be able to be enforced and managed properly in softwarized and virtualized networks (SDN/NFV) and CPS/IoT systems. In addition, these security enablers and protocols need to be redesigned to cope with the constrained nature of distributed IoT networks, that requires lightweight crypto-protocols and solutions to be enforced in constrained (battery, memory, cpu) devices and networks.

○ *Security in 5G mMTC and mobile networks*: 5G mMTC (massive Machine-type Communications) is the key technology needed to scale up the internet of thing (IoT). However, this 5G large-scale management and orchestration raises new cybersecurity threats which requires novel security solutions, as analysed in [26]. 5G imports vulnerabilities and threats coming from cloud computing, virtualization and SDN/NFV technologies. Thus, it is a research challenge to deal with information transmission management, secure communication channels, new security interfaces for AAA to deal with Non-Access Spectrum (NAS) signalling, roaming security, and cope with diverse network-based mobile security threats and attacks (e.g. saturation attacks, penetration attacks, identity thief, Man-in-the-middle, scanning attaks, Hijacking, DoS attacks, Signaling storms).

3. Cognitive detection and mitigation of evolving new kind of cyber-threats

o *Dealing with evolving kind of cyberattacks*: The identification of novel types of attacks not yet identified before (e.g. unknown zero-day attacks), that can exploit IoT networks, CPS (and the consequent protection approaches to provide advanced security from last generation threats) is a key research challenge. This new kind of attacks need to be addressed following a global approach through both, signature-based and anomaly-based detection techniques, by using artificial intelligence and Big Data analysis approaches. In the cyber physical world, the attacker's goal is to disrupt both the normal operations of the CPS, e.g. sensor readings, safety limits violation, status reports, safety compliance violation etc. and communication flows among devices. The continued rise of cyber-attacks together with the evolving skills of the attackers, and inefficiency of the traditional security algorithms to defend against advanced and sophisticated attacks such as DDoS, slow DoS and zero-day, demand the development of novel defence and resilient detection techniques.

o *Monitoring in heterogenous ICT systems.* Cybersecurity handling, especially in Critical systems, Cyber Physical Systems and IoT networks introduces challenges due the restrictions and constrained nature of these kind of devices and networks. New tools, for network scanning (including encrypted traffic), analysis of digital forensics and pen testing as well as innovative algorithms and techniques (e.g. machine learning) are needed to perform security analysis.

o *Real-time incident detection and analysis*: Incident analysis should be supported by risk models that follows a multidimensional approach, performing evaluation of incidents that combines several factors (such as, for instance, incident severity, criticality of assets affected, global risk associated to the incident or cost of potential mitigations among others) to decide, if needed, dynamically the most convenient mitigation plan to enforce. It should cover, threat analysis, data fusion and correlation from different sources different types of events to detect hidden relations and thus identify potential threats.

o *Cyber situational-awareness, self-learning and dynamic reaction for self-healing, self-repair and self-protection capabilities*: Management and Control systems as well as Autonomous systems, such as for instance, drones, smart objects, self-driving cars, robots, etc, will need to perform self-learning to make proper intelligent decisions based on current real-time situation. However, those autonomous systems could be manipulated when sensing the external world, and therefore, assessing the quality of the potential sensed environment is a challenge. In addition, upcoming cybersecurity frameworks and systems should face the challenge of countering dynamically cyberattacks according to contextual and evolving conditions, thereby providing self-healing, self-repair and self-protection capabilities. This will allow to diagnose and enforce proper defence mechanism and mitigate threats autonomously.

○ *Cognitive big data analysis of systems/networks, services, social networks and cybersecurity intelligence information to counter cyberthreats*: To meet this challenge an interdisciplinary approach should be followed, performing cognitive science, communications, computational linguistics, discourse processing, language studies and social psychology. Upcoming cybersecurity solutions should meet the challenge of combing diverse technologies, such for instance, IA algorithms, Machine Learning (ML), CEP (Complex Event Processing), SNA (Social Network Analysis) and NLP (Natural Language processing) to assess systems data/events, social features in communications used by terrorist organizations, in order to increase security levels and counter cyber-threats.

4. **Dynamic risk assessment and evaluation of cybersecurity, trustworthiness levels and legal compliance of ICT systems**
New models are needed to quantify in real time, according to the context, the trustworthiness, of new kind of devices-system-networks, compute the risk associated to an ICT system and evaluate the security and privacy legal compliance. Risk evaluation should be performed through an interdisciplinary approach including not only technological, but also legal and socio-ethical perspectives. Relevant metrics need to be established for cybersecurity economic analysis, cybersecurity and cybercrime market. The risk evaluation should consider automated analysis, for behavioural, social analysis, cybersecurity risk and cost assessment. In this regard, another challenge is to make this risk analysis usable and easy interpretable for administrators and stakeholders, through short and long terms actions and recommendations.

Another related challenge is to kept users informed about the trustworthiness levels of their application and servers, according to multi factor criteria, encompassing sociocultural, legal, ethical, technological and business while paying due attention to the protection of Human Rights. Proper recommendations about certification and labelling of ICT products and services should be automatically inferred, that will foster trust among citizens that use them.

5. **Digital forensics handling, security intelligence and incident information exchange**

   An important cybersecurity challenge is to improve levels of collaboration between cooperative and regulatory approaches for information sharing in order to enhance cybersecurity and mitigate the risk and the impact of cyber-attacks. In this regard, new standards, models, protocols are needed to achieve interoperability for effective collaboration between operational teams including Law Enforcement Agencies, CSIRTs, Organization, through automated exchange of cyber-crime data, including source Open Source Intelligence (OSINT) data sources, thereby allowing sharing the own system cyber-situational awareness information with the external entities in an effective way. In addition, another challenge is to perform automatic application and enforcement of data sharing in an interoperable manner that can feed the incident analysis, which ultimately, can help in the cybersecurity decision support making.

6. **Cybersecurity and privacy tools for end-users and SMEs. The usability and human factor challenges**

   Individuals, SMEs, local administrators and related end-users are overwhelmed with the complexity of cybersecurity and privacy aspects, which obstructs proper decision making and digital technology usage. These kinds of users cannot dedicate enough effort and resources to invest in security personnel and cybersecurity products or services. User-friendly and automated cybersecurity unified tools need to implemented targeting (potential inexpert) final users, so that they can face cybersecurity threats and manage properly security configurations. The human factor is one of the most problems when it comes to security management, as it can easily generate new security gaps. Most of the cyber-attacks such as ramsonware, physing, identity chief, etc, are originated by the end-user. Thus, the human factor needs to be handled by cybersecurity frameworks and tools in order to increase system resilience against end-users' and operators' errors.

1.2.2 Privacy and Trust Related Research Challenges

7. **Reliable and privacy-preserving physical and virtual identity management**

   Identity management Systems require new security and privacy mechanisms that can holistically manage user's/object's privacy, ID-proofing techniques based on multiple biometrics, strong authentication, usage of breeder documents (e.g. eID, ePassports), while ensuring privacy-by-default, unlikability, anonymity, federation support, non-reputation and self-sovereign IdM management. The challenge is to manage properly those features for mobile, online or physical/face-to face scenarios, while maintaining usability and compliance with regulation e.g. GDPR (General Data Protection Regulations)[GDPR] and eIDAS [21]. This will allow ultimately to reduce identity-theft and related cybercrimes.

   In this context, another challenge arises from the extension of global identity management and AAA to *anything* deployments, managing efficiently identities and access control of new kinds of autonomous Systems, such as, IoT smart objects, self-driving cars, robots, humanoids, drones, etc. that requires new evolved algorithms, protocols and systems.

8. **Efficient and secure cryptographic mechanisms to strengthen confidentiality and privacy**

   ○ *Confidentiality and privacy in distributed systems*: End-to-end encryption of shared data, in transit and in rest, while maintaining usability and efficiency on the end-user side is an open research challenge that still needs to be covered effectively to protect user's privacy. In this sense, new techniques, algorithms and protocols, e.g. those based on proxy re-encryption, are needed to reinforce security/privacy while outsourcing the computation to Cloud wallets to minimize user's risks in protecting crypto-material. In addition, new crypto-privacy techniques are needed to guarantee authenticity on the data through novel signatures schemes.

   ○ *Data anonymization and secure data sharing*: All exchanged data should be encrypted, without intermediate entities such as proxies or cloud-providers being able to access the user's data. Data minimization and privacy-by-default properties, above all, in emerging distributed deployments needs to be guaranteed. Thus, novel crypto-privacy protocols, mechanism and systems, such as those based on Zero-knowledge proofs, are needed to ensure anonymity, minimal disclosure of personal information, above all in public Clouds, ledgers and mobiles, while ensuring the user's rights laid out in GDPR.

o *Big data privacy*: Data analytics raises new concerns about privacy preservation, as the possible dynamic combination of large data coming from diverse sources can undermine anonymity, pseudonimity properties that can be given for granted in a single domain. This challenge is especially relevant in critical sectors (eHealth, eBanking), distributed systems that will handle massive user data, e.g. blockchains, ledgers, and social networks. Therefore, new technologies to enforce efficient privacy protection are needed, as a response of a new collaborative privacy-assessment mechanisms.

o *Crypto-resilience to brute-force attacks*: Quantum computing technology is making possible new risks and threats, as most of current encryption and signature algorithms will not be fully secure against brute-force attacks perpetrated by quantum computers. In this sense, new cryptographic algorithms are needed to be resilient to brute-force attacks using quantum computing.

9. **Global trust management of eID and related services**
There is a need of a Global, trusted, open and scalable infrastructure where authorities can publish their trust information to certify trustworthy electronic identities, so that rest of stakeholders, including public sector, private companies, and citizens can verify automatically trust in electronic transactions, while hiding the complexity of dealing with heterogenous formats and protocols.

This challenging Global Trust System should deal with issues such as unified data model, rights delegation, trust policy language, claims discovery to make the system interoperable accessible for everyone, while facilitating, at the same time, the use of eID and electronic signature technology in real world applications. This global trust management infrastructure should leverage the eIDAS trust scheme laid out in Regulation (EU) N°910/2014 [21], extending the European Trust Service Status List (TSL) infrastructure towards a "Global Trust List".

10. **Privacy assessment, run-time evaluation of the quality of security and privacy risks** There is a need of evaluation tools and methods to assess whether an application or a service is compliant with privacy and personal data protection principles, as well as quantitative and qualitative run-time evaluation of the quality of security and privacy risks.

In this sense, novel Dynamic Security and Privacy Seals (DSPS) are needed to increase trust in the system, by combining ISO, legal norms and security and privacy standards with deep technical monitoring integration, in order to provide a user-friendly and synthetic view of the overall system trust ability. In this regard, it is challenging to integrate and enhance the alerts generated by the underlying systems with direct technical and organizational feedback from the end-user. These novel kinds of seals would come up with legally valid and non-repudiable proof of compliance of the system with legal or contractual security-privacy requirements, which can be easily managed and visualized by the user.

1.3 H2020 Projects Facing the Challenges

1.3.1 Cybersecurity Related Projects Addressing the Challenges

- **ANASTACIA** [5] (Chapter 02): ANASTACIA is researching, developing and demonstrating a holistic solution enabling trust and security by-design for Cyber Physical Systems (CPS) based on IoT and Cloud architectures. ANASTACIA cybersecurity framework provides self-protection, self-healing and self-repair capabilities through novel enablers and components. The framework dynamically orchestrates and deploys security policies and actions that can be instantiated on local agents. Thus, security is enforced in different kinds of devices and heterogeneous networks, e.g. IoT – or SDN/NFV – based networks. The framework has been designed in full compliance to SDN/NFV standards as specified by ETSI NFV and OFN SDN, respectively. Therefore, Anastacia is addressing challenges #1, #2, #3 and #4 enumerated in Section 2.1

- **SAINT** [6] (Chapter 03): "SAINT analyses and identifies incentives to improve levels of collaboration between cooperative and regulatory approaches to information sharing. SAINT is designing new methodologies for the development of an ongoing and searchable public database of cybersecurity indicators and open source intelligence. Comparative analysis of cyber-crime victims and stakeholders within a framework of qualitative social science methodologies deliver valuable evidences and advance knowledge on privacy issues and deep web practices. SAINT defines innovative models, algorithms and automated framework for cost-benefit analysis and estimation of tangible and intangible costs for optimal risk and investment incentives". Thus, SAINT is mainly focusing on challenge #5 enumerated in Section 2.1.

- **FORTIKA** [7] (Chapter 04): "The project is designing and implementing a security 'seal' specially devised for small and medium-sized companies that will strengthen trust and facilitate further adoption of digital technologies. The project is implementing robust, resilient and effective cybersecurity solutions to be customized for each individual enterprise's evolving needs and can also speedily adapt/respond to the changing cyber threat landscape". Therefore, FORTIKA is mainly focusing on challenges #2 and #6 of those described in Section 2.1.

- **CYBECO** [8] (Chapter 05): "CYBECO focuses on two mains aspects to deal with cyber-insurance from a Behavioural Choice Perspective: (1) including cyber threat behaviour through adversarial risk analysis to support insurance companies in estimating risks and setting premiums and (2) using behavioural experiments to improve IT owners' cybersecurity decisions. Therefore, CYBECO facilitates risk-based cybersecurity investments supporting insurers in their cyber offerings through a risk management modelling framework and tool." Therefore, SAINT is mainly focusing on challenge #4 of Section 2.1.

- **SISSDEN** [9] (Chapter 06): "SISSDEN is intended to improve the cyber security through development of situational awareness and sharing of actionable information. The passive threat data collection mechanism is complemented by behavioural analysis of malware and multiple external data sources. Actionable information produced by SISSDEN provides no-cost victim notification and remediation via organizations such as CERTs, ISPs, hosting providers and LEAs such as EC3. The main goal of the project is the creation of multiple high-quality feeds of actionable security information that can be used for remediation purposes and for proactive tightening of computer defences. This is achieved through the development and deployment of a distributed sensor network based on state-of-the-art honeypot and darknet technologies, the creation of a high-throughput data processing centre, and provisioning of in-depth analytics, metrics and reference datasets of the collected data." Therefore, SISSDEN is mainly focusing on challenge #5 of Section 2.1.

- **CIPSEC** [10] (Chapter 07): "CIPSEC aims to create a unified security framework that orchestrates state-of-the-art heterogeneous security products to offer high levels of protection in IT (information technology) and OT (operational technology) departments of CIs, also offering a complete security ecosystem of additional services. These services include vulnerability tests and recommendations, key personnel training courses, public-private partnerships (PPPs), forensics analysis, standardization activities and analysis against cascading effects." CIPSEC is mainly focusing on challenge #3, #4 and #5 of Section 2.1.

- **CS-AWARE** [11] (Chapter 08): CS-AWARE aims to increase the automation of cybersecurity awareness approaches, by collecting cybersecurity relevant information from sources both inside and outside of monitored local public administrations (LPA) systems, performing advanced big data analysis to set this information in context for detecting and classifying threats and to detect relevant mitigation or prevention strategies. CS-AWARE aims to advance the function of a classical decision support system by enabling supervised system self-healing in cases where clear mitigation or prevention strategies for a specific threat could be detected. CS-AWARE is built around this concept and relies on cybersecurity information being shared by relevant authorities in order to enhance awareness capabilities. At the same time, CS-AWARE enables system operators to share incidents with relevant authorities to help protect the larger community from similar incidents. CS-AWARE is mainly focusing on challenge #5 of Section 2.1.

- **RED-Alert** [12] (Chapter 09): "RED-Alert has built a complete software toolkit to support LEAs in the fight against the use of social media by terrorist organizations for conducting online propaganda, fundraising, recruitment and mobilization of members, planning and coordination of actions, as well as data manipulation and misinformation. The project aims to cover a wide range of social media channels used by terrorist groups to disseminate their content which will be analysed by the RED-Alert solution to support LEAs to take coordinated action in real time but having as a primordial condition preserving the privacy of citizens." RED-Alert is mainly focusing on challenge #3 of Section 2.1.

- **Truessec.eu** [13] (Chapter 10): "The main goal of TRUESSEC project is to foster trust and confidence in new and emerging ICT products and services throughout Europe by encouraging the use of assurance and certification processes that consider multidisciplinary aspects such as sociocultural, legal, ethical, technological and business while paying due attention to the protection of Human Rights." Therefore, TRUESSEC is mainly addressing challenge #4.

Table 1.1   Main cybersecurity research challenges and related EU project's

Challenge ID	Name	EU projects addressing the challenge
1	Interoperable and scalable security management in heterogeneous ecosystems	ANASTACIA
2	Autonomic Security orchestration and enforcement in softwarized and virtualized IoT/CPS systems and mobile environments	ANASTACIA, FORTIKA, CIPSEC
3	Cognitive detection and mitigation of evolving new kind of cyber-threats	ANASTACIA, CIPSEC, CS-AWARE, RED-ALERT
4	Dynamic Risk assessment and evaluation of cybersecurity, trustworthiness levels, privacy and legal compliance of ICT systems	CYBECO, CIPSEC, TRUESSEC, ANASTACIA
5	Digital Forensics handling, security intelligent and incident information exchange	SIESSDEN, SAINT, CIPSEC, CS-AWARE
6	Cybersecurity and privacy tools for end-users and SMEs. The usability and human factor challenges	FORTIKA

Table 1.1 recaps the main cybersecurity research challenges presented in Section 1.2.1 and links them with the EU project's, presented in this section, that are addressing those challenges.

1.3.2 H2020 Projects Addressing the Privacy and Trust Related Challenges

- **ARIES** [14] (Chapter 11): Aries aims to set up a reliable identity ecosystem encompassing technologies, processes and security features that ensure highest levels of quality in secure credentials for highly secure and privacy-respecting physical and virtual identity management processes with the specific aim to tangibly achieve a reduction in levels of identity fraud, theft, wrong identity and associated crimes. The ecosystem is strengthening the link between physical documents linked to the biometric identity and the digital (online and mobile) identity.

- **LIGHTest** [15] (Chapter 12): LIGHTest project aims to set-up a global trust infrastructure where authorities can publish their trust information. Thus, member states can use infrastructure to publish lists of qualified trust services, while private companies can establish trust in different sectors, such as, inter-banking, international trade, shipping, business reputation and credit rating. Then, different entities can query this trust information to verify trust in simple signed documents or multi-faceted complex transactions.

- **CREDENTIAL** [16] (Chapter 13): CREDENTIAL project has developed a cloud-based service for identity provisioning and data sharing. On the one hand, it offers high confidentiality and privacy guarantees to the data owner, while, on the other hand, it offers high authenticity guarantees to the receiver. CREDENTIAL integrates advanced cryptographic mechanisms into standardized authentication protocols. The solution has proved high user convenience, strong security, and practical efficiency.
- **FutureTrust** [17] (Chapter 14): The FutureTrust project aims to develop a comprehensive Open Source validation service as well as a scalable preservation service for electronic signatures and will provide components for the eID-based application for qualified certificates across borders, and for the trustworthy creation of remote signatures and seals in a mobile environment. Furthermore, the FutureTrust project extends and generalize existing trust management concepts to build a "Global Trust List", which allows to maintain trust anchors and metadata for trust services and eID related services around the globe.
- **LEPS** [18] (Chapter 15): LEPS project aims to "validate and facilitate the connectivity options to recently established eIDAS ecosystem, which provides this trusted environment with legal, organisational and technical guarantees already in place. Strategies have been devised to reduce SP implementation costs for this connectivity to eIDAS technical infrastructure". The project has implemented integrated and validated the solution in Pilots of two EU countries.

Table 1.2 summarizes the main privacy-related research challenges presented in Section 1.2.2 and links them with the EU project's, presented in this section, that are addressing those challenges.

Table 1.2 Main Privacy related research challenges and related EU projects

Challenge ID	Name	EU projects addressing the challenge
7	Reliable and privacy-preserving physical and virtual identity management	ARIES, LEPS
8	Efficient and secure cryptographic mechanisms to strengthen confidentiality and privacy	CREDENTIAL
9	Global trust management of eID and related services	LIGHTest, Future Trust
10	Privacy assessment, run-time evaluation of the quality of security and privacy risks	ANASTACIA

1.4 Conclusion

This chapter has identified and introduced the 10 main cybersecurity and privacy research challenges presented and addressed in this book by 14 European research projects. Some of the challenges revolve around the autonomic cybersecurity management, orchestration and enforcement in heterogeneous and virtualized CPS/IoT and mobile ecosystems. The challenges identified cognitive detection and mitigation of evolving new kind of cyber-threats; the dynamic risk assessment and evaluation of cybersecurity, trustworthiness levels, privacy and legal compliance of ICT systems; the digital Forensics handling; the security intelligent and incident information exchange; and cybersecurity and privacy tools and the associated usability and human factor. Regarding privacy and trust related challenges, we have identified four main global ones, encompassing the reliable and privacy-preserving identity management, efficient and secure cryptographic mechanisms, Global trust management and privacy assessment.

In addition, the chapter has introduced the 14 EU projects analysed in the book and the main challenges the are addressing. ANASTACIA, SAINT, FORTIKA, CYBECO, SISSDEN, CIPSEC, CS-AWARE. RED-Alert, Truessec.eu. ARIES, LIGHTest, CREDENTIAL, FutureTrust.

The rest of the book is intended to present each of those 14 EU projects, which are described in a different book chapter. Each chapter includes the project's overviews and objectives, the particular challenges they are covering, research achievements on security and privacy, as well as the techniques, outcomes, and evaluations accomplished in the scope of the EU project.

Acknowlegdements

This work has been supported by a postdoctoral INCIBE grant "Ayudas para la Excelencia de los Equipos de Investigación Avanzada en Ciberseguridad" Program, with Code INCIBEI-2015-27363. This book chapter has also received funding from the European Union's Horizon 2020 research and innovation programme under grant agreement No. 700085 (ARIES project).

References

[1] Cybersecurity Strategy of the European Union: An Open, Safe and Secure Cyberspace. Joint communication to the European Parliament, the Council, the European Economic and Social Committee and the Committee of the Regions. (2013). Available at from: https://eeas.europa.eu/archives/docs/policies/eu-cyber-security/cybsec _ comm_en.pdf

[2] H2020-EU.3.7. – Secure societies – Protecting freedom and security of Europe and its citizens. https://cordis.europa.eu/programme/ rcn/664463/en

[3] Secure societies – Protecting freedom and security of Europe and its citizens Last accessed 10/04/1 from: https://ec.europa.eu/programmes/hori zon2020/en/h2020-section/secure-societies-%E2%80%93-protecting-fr eedom-and-security-europe-and-its-citizens

[4] European projects Clustering workshop On Cybersecurity and Privacy (ECoSP 2018) https://2018.ares-conference.eu/workshops/ecosp-2018/. held in conjunction with the 13th International Conference on Availability, Reliability and Security (ARES 2018 – http://www.ares-conference.eu)

[5] ANASTACIA (Advanced Networked Agents for Security and Trust Assessment in CPS / IOT Architectures) H2020 EU project, Grant Agreement No. 731558 http://anastacia-h2020.eu/

[6] SAINT (Systemic Analyzer In Network Threats) H2020 EU project, Grant Agreement No. 740829 https://project-saint.eu/

[7] FORTIKA (Cyber Security Accelerator for trusted SMEs IT Ecosystem) H2020 EU project, Grant Agreement No. 740690. http://fortika-project.eu/

[8] CYBECO (Supporting Cyberinsurance from a Behavioural Choice Perspective) H2020 EU project, Grant Agreement No. 740920. https://www.cybeco.eu/

[9] SISSDEN (Secure Information Sharing Sensor Delivery Event Network) H2020 EU project, grant Agreement No. 700176. https://sissden.eu/

[10] CIPSEC (Enhancing Critical Infrastructure Protection with innovative SECurity framework) H2020 EU project, Grant Agreement No. 700378 http://www.cipsec.eu/

[11] CS-AWARE (A cybersecurity situational awareness and information sharing solution for local public administrations based on advanced

big data analysis) H2020 EU project, Grant Agreement No. 740723. https://cs-aware.eu/

[12] RED-Alert (Real-time Early Detection and Alert System) H2020 EU project, Grant Agreement No. 740688 http://redalertproject.eu/

[13] Truessec.eu (TRUst-Enhancing certified Solutions for SEcurity and protection of Citizens' rights in digital Europe) H2020 EU project, Grant Agreement No. 731711 http://truessec.eu/

[14] Aries (ReliAble euRopean Identity EcoSystem), H2020 EU Project Grant Agreement No. 700085 https://www.aries-project.eu/

[15] LIGHTest (Lightweight Infrastructure for Global Heterogeneous Trust management in support of an open Ecosystem of Stakeholders and Trust schemes), H2020 EU Project Grant Agreement No. 700321, https://www.lightest.eu/

[16] CREDENTIAL (Secure Cloud Identity Wallet), H2020 EU project, Grant Agreement No. 653454, https://credential.eu/

[17] FutureTrust (Future Trust Services for Trustworthy Global Transactions), H2020 EU project, Grant Agreement No. 700542 https://www.futuretrust.eu/

[18] LEPS (Leveraging eID in the Private Sector), European Union's Connecting Europe Facility, Grant Agreement No. INEA/OEF/ICT/A2016/1271348. http://www.leps-project.eu/

[19] Roman, R., Zhou, J., & Lopez, J. (2013). On the features and challenges of security and privacy in distributed internet of things. Computer Networks, 57(10), 2266–2279.

[20] Zou, Y., Zhu, J., Wang, X., & Hanzo, L. (2016). A survey on wireless security: Technical challenges, recent advances, and future trends. *Proceedings of the IEEE*, 104(9), 1727–1765.

[21] European Parliament, 'Regulation (EU) No. 910/2014 of the European Parliament and of the Council of 23 July 2014 on electronic identification and trust services for electronic transactions in the internal market and repealing Directive 1999/93/EC', European Parliament, Brussels, Belgium, Regulation 910/2014, 2014.

[22] Ziegler, S., Crettaz, C., Kim, E., Skarmeta, A., Bernabe, J. B., Trapero, R., & Bianchi, S. (2019). Privacy and Security Threats on the Internet of Things. In *Internet of Things Security and Data Protection* (pp. 9–43). Springer, Cham.

[23] Megatreds (2018). "Study on global megatrends in cybersecurity, ponemon institute research report", Research report, February 2018.

[24] Backes, M., Buxmann, P., Eckert, C., Holz, T., Müller-Quade, J., Raabe, O., & Waidner, M. (2016). *Key Challenges in IT Security Research.* Discussion Paper for the Dialogue on IT Security 2016, SecUnity, https://it-security-map. eu.

[25] Lu, Y., Da Xu, L. Internet of Things (IoT) cybersecurity research: a review of current research topics. *IEEE Internet of Things Journal*, 2018.

[26] Ahmad, I., Kumar, T., Liyanage, M., Okwuibe, J., Ylianttila, M., & Gurtov, A. (2017, September). 5G security: Analysis of threats and solutions. In *2017 IEEE Conference on Standards for Communications and Networking (CSCN)* (pp. 193–199). IEEE.

2

Key Innovations in ANASTACIA: Advanced Networked Agents for Security and Trust Assessment in CPS/IOT Architectures

Jorge Bernal Bernabe[1], Alejandro Molina[1], Antonio Skarmeta[1], Stefano Bianchi[2], Enrico Cambiaso[3], Ivan Vaccari[3], Silvia Scaglione[3], Maurizio Aiello[3], Rubén Trapero[4], Mathieu Bouet[5], Dallal Belabed[5], Miloud Bagaa[6], Rami Addad[6], Tarik Taleb[6], Diego Rivera[7], Alie El-Din Mady[8], Adrian Quesada Rodriguez[9], Cédric Crettaz[9], Sébastien Ziegler[9], Eunah Kim[10], Matteo Filipponi[10], Bojana Bajic[11], Dan Garcia-Carrillo[12] and Rafael Marin-Perez[12]

[1]Department of Information and Communications Engineering, University of Murcia, Murcia, Spain
[2]Research & Innovation Department, SOFTECO SISMAT SRL, Di Francia 1 – WTC Tower, 16149, Genoa, Italy
[3]National Research Council (CNR-IEIIT) – Via De Marini 6 – 16149 Genoa, Italy
[4]Atos Research and Innovation, Atos, Calle Albarracin 25, Madrid, Spain
[5]THALES Communications & Security SAS, Gennevilliers, France
[6]Department of Communications and Networking, School of Electrical Engineering, Aalto University, Finland
[7]R&D Department, Montimage, 75013, Paris, France
[8]United Technologies Research Center, Ireland
[9]Mandat International, Research, Switzerland
[10]Device Gateway SA, Research and Development, Switzerland
[11]Archimede Solutions, Geneva, Switzerland
[12]Department of Research & Innovation, Odin Solutions, Murcia, Spain
E-mail: jorgebernal@um.es; alejandro.mzarca@um.es; skarmeta@um.es; stefano.bianchi@softeco.it; enrico.cambiaso@ieiit.cnr.it; ivan.vaccari@ieiit.cnr.it; silvia.scaglione@ieiit.cnr.it; maurizio.mongelli@ieiit.cnr.it; ruben.trapero@atos.net;

mathieu.bouet@thalesgroup.com; dallal.belabed@thalesgroup.com;
miloud.bagaa@aalto.fi; rami.addad@aalto.fi; tarik.taleb@aalto.fi;
diego.rivera@montimage.com; madyaa@utrc.utc.com;
aquesada@mandint.org; ccrettaz@mandint.org; sziegler@mandint.org;
eunah.kim@devicegateway.com; mfilipponi@devicegateway.com;
bbajic@archimede.ch; dgarcia@odins.es; rmarin@odins.es

This book chapter presents the main key innovations being devised, implemented and validated in the scope of Anastacia H2020 EU research project, to meet the cybersecurity challenge of protecting dynamically heterogenous IoT scenarios, endowed with SDN/NFV capabilities, which face evolving kind of cyber-attacks. The key innovations encompasses, among others, policy-based security management in IoT networks, trusted and dynamic security orchestration of virtual networks security functions using SDN/NFV technologies, security monitoring and cognitive reaction to countering cyber-treats, behavioural analysis, anomaly detection and automated testing for the detection of known and unknown vulnerabilities in both physical and virtual environments as well as secured and authenticated dynamic seal system as a service.

2.1 Introduction

The Internet of Things (IoT) aims to leverage network capabilities of devices and smart objects, integrating the sensing and actuation features to create pervasive information systems, which are used as baseline to provide smart services to the industry and citizens. However, as a greater number of constrained IoT devices are connected to Internet, the security and privacy risks increase accordingly. The boosted connectivity and constrained capabilities of devices in terms of memory, CPU, memory, battery, the unattended behaviour of IoT devices, misconfigurations and lack of vendor support, increase potential kinds of vulnerabilities. Therefore, new advanced security frameworks for IoT deployments are needed to face these threats and meet dynamically the desired defence levels.

H2020 Anastacia EU project addresses the security management of heterogenous and distributed IoT scenarios, such as Smart Buildings or Smart Cities, which can benefit from a policy-based orchestration and security management approach, where NFV/SDN-based solutions and novel monitoring and reaction tools are combined to deal with new kind of evolving cyber-attacks.

ANASTACIA is developing new methodologies, frameworks and support tools that will offer resilience to distributed smart IoT systems and Mobile Edge Computing (MEC) scenarios against cyber-attacks, by leveraging SDN and NFV technologies. Security VNFs can be timely and dynamically orchestrated through policies to deal with heterogeneity demanded by these distributed IoT deployments that can be deployed either at the core of at the edge, in VNF entities, to rule the security in IoT networks. Dynamic and reactive provisioning of Security VNFs towards the edge of the network can enhance scalability, necessary to deal with IoT scenarios.

The primary objective of the ANASTACIA project is to address cyber-security concerns by researching, developing and demonstrating a holistic solution enabling trust and security by-design for Cyber Physical Systems (CPS) based on Internet of Things (IoT) and Cloud architectures.

The heterogeneous, distributed and dynamically evolving nature of CPS based on IoT and virtualised cloud architectures introduces new and unexpected risks that can be only partially solved by current state-of-the-art security solutions. Innovative paradigms and methods are required i) to build security into the ICT system at the outset, ii) to adapt to changing security conditions, iii) to reduce the need to fix flaws after deploying the system, and iv) to provide the assurance that the ICT system is secure and trustworthy at all times. ANASTACIA is thus developing, integrating and validating a security and privacy framework that will be able to take autonomous decisions through the use of new networking technologies such as Software Defined Networking (SDN) and Network Function Virtualisation (NFV) and intelligent and dynamic security enforcement and monitoring methodologies and tools.

Dealing with this general ambition and scenario raises several research challenges, being faced in Anastacia:

- Interoperable and scalable IoT security management: dealing with the level of abstraction, the language and new security models, contextual IoT aspects in policies, particularities in IoT security models, policy conflicts and dependencies in orchestration policies.
- Optimal selection of SDN/NFV-based security mechanisms: allocate multiple VNF requests on an NFV Infrastructure, especially in a cost-driven objective.
- Orchestration of SDN/NFV-based security solutions for IoT environments: the selection of the adequate mitigation plan and the fast

enforcement of the defined policies, as well as orchestration and the enforcement of the adequate countermeasures in a short time.

- Dealing with a new kind of cyber-attacks in IoT: providing advanced security from last generation threats on IoT environments.
- Learning decision model for detecting malicious activities: the development of novel defence and resilient detection techniques.
- Hybrid security monitoring for IoT enhanced with event correlation: The application of both signature-based and behavioural-based security analysis for IoT.
- Quantitative evaluation of incidents for mitigation support: combination of several factors to evaluate incidents to decide on the most convenient mitigation plan to enforce.
- Construction of a dynamic security and privacy seal that secures both organizational and technical data: generate trust by considering technical insights on security and privacy personal data protection requirements.

This chapter describes the main key innovations being devised, implemented and evaluated in the scope of ANASTACIA to cope with the aforementioned security challenges in IoT scenarios.

2.2 The Anastacia Approach

2.2.1 Anastacia Architecture Overview

The NIST Cybersecurity Framework identifies five steps for the protection of critical infrastructures: Identification, protection, detection, response and recovering. In general, these three steps are supported by the retrieval and management of security information extracted from the infrastructure to protect. On top of the five steps of the NIST Cybersecurity Framework, we can overlap the three main activities in what regards to the data lifecycle in ICT infrastructures for security protection, namely the data acquisition, data dissemination, data consumption and data processing. Data acquisition includes of the components and mechanisms to retrieve relevant data from the infrastructure, such as logs, heartbeats or reports. Data Dissemination regards to the elements that allow to distribute or store the acquired data among the relevant components of the infrastructure, such as monitoring agents, document or software repositories. Data consumption refers to the components involved in the usage of such data, either for its correlation, patterns finding for incident detection or forensic analysis. Finally, data processing carries out activities based on the result obtained by the data consumers, such as mitigation actions to react to the incidents detected, their enforcement or the creation of security and privacy seals that inform about the security and privacy level of the platform.

The ANASTACIA approach is based on the flow and management of data gathered from IoT infrastructures. Following the aforementioned model, ANASTACIA designs and uses proper mechanisms to retrieve information from the underlying infrastructure and accurate ways to interpret it them to know the real status of the infrastructure and to make accurate decisions based to automatically react to incidents. ANASTACIA relies on the concept of automation when referring to the dynamic protection against security incidents, considering the cycle depicted in Figure 2.1 for identifying sources of relevant security information, deployment of security probes for the protection of IoT infrastructures, the detection of security incidents, responding to them by generating security alerts that are used to enforce mitigation actions to recover from the detected security incidents.

To this end ANASTACIA has designed a plane-based architecture [1] where the information flows from the data acquisition from the IoT infrastructure to their dissemination and consumption by the monitoring infrastructure and to the data processing by the reaction module to decide about mitigations to enforce. Figure 2.2 represents the plane-based approach of ANASTACIA. On top of the data plane, which represents the data to obtain from the IoT infrastructure, and on top of the control plane, which represents the

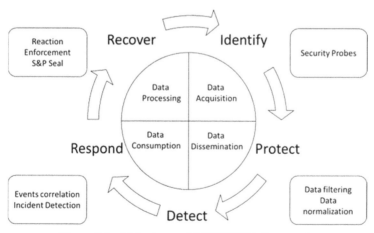

Figure 2.1 Main stages of ANASTACIA framework.

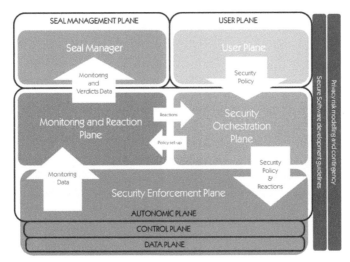

Figure 2.2 Anastacia high-level architectural view.

elements (software defined networks or virtual network functions) that allows to interact with the IoT infrastructure, are: (i) the enforcement plane that uses the control plane to obtain monitoring data from the infrastructure, (ii) the monitoring and reaction plane, which correlates the monitoring data to detect incidents and propose reactions to mitigate them, (iii) the security orchestration plane, which enforce the reactions using the enforcement plane. On top of them, the Seal Management plane uses monitoring data and reactions to provide with a snapshot of the security and privacy level of the infrastructure, and the user plane that provides interaction with human administrators for the establishment of security policies.

2.3 Anastacia Main Innovation Processes

2.3.1 Holistic Policy-based Security Management and Orchestration in IOT

In distributed smart IoT deployments scenarios like those previously described, the system security management is crucial. At this point, it is important to highlight that to the diversity of the current systems and services they are added a vast amount of different devices in the IoT domain, being the latter quite different among the previous approach and even among themselves. From this point of view, the current state of art shows that it is highly valuable to provide different levels of security policies to provide different levels of abstraction for different profiles of management. It is also important to highlight the difference between generic models and specific extensible models, as well as to remark then relevance of policy orchestration features and policy conflict detection. Main ANASTACIA's contributions on policies reside in the unification of relevant, new and extended capability-based security policy models (including ECA features), as well as policy orchestration and conflict detection mechanisms, all under a unique policy framework. To this aim, the holistic policy-based solution provides different components and features like **Policy Models**, **Policy Editor Tool**, **Policy Repository**, **Policy Interpreter**, **Policy Conflict Detection** and **Policy for Orchestration**.

ANASTACIA's **Policy Models** thus improve the current state of the art as well as provide novelty approaches to be able to increase the security measures and countermeasures in the whole system at different levels. To this aim, ANASTACIA adopts and extend concepts and features from the state of art, to provide a unified security policy framework. I.e., ANASTACIA involves and evolves previous works by extending the already existing features as well as by providing new IoT-focused features.

The Policy Models can be instantiated using the **Policy Editor Tool** which allows defining security policies at a high-level of abstraction through a friendly GUI. In this way, the security administrator is able to manage the security of the system by instantiating new security policies, as well as supervise the existing security policies by the Policy Repository. **The Policy Repository** registers all policy operations as well as the current status for each one. It also provides valuable policy templates to make the security management easier.

Since the security policies are instantiated in a High-level Security Policy Language (HSPL), it must be transformed in configurations for the specific devices which will enforce the security policy. To this aim, the **Policy Interpreter** is able to refine the HSPL in one or several Medium-level Security Policy Language (MSPL) policies depending on a set of identified capabilities (filtering, forwarding, etc.). This process transforms the high-level concepts into more detailed parameters but still independent to the specific technologies. Finally, these MSPL policies are translated in final configurations using specific translator plugins for each technology. Once the configurations have been obtained, they can be enforced in the specific security enablers, understanding a security enabler as a piece of hardware or software able to implement a specific capability. Of course, a security policy only can be enforced if it does not present any kind of conflict with the already enforced ones. In this sense the **Policy Conflict Detection** engine verifies that the new security policy will not generate conflicts like redundancy, priorities, duties (e.g. packet inspection vs channel protection), dependences or contradictions. To this aim, the security policy is processed against the rule engine which extracts context information from the policy repository and the system model to perform the necessary verifications.

Regarding the dependences, ANASTACIA also includes as part of the policy model the Policy for Orchestration concept. The **Policy for Orchestration** model allows the security administrator to specify how a set of security policies must be enforced by defining priorities and dependencies, where a security policy can depend on other security policies or even in system events like an authentication success.

Through these components and features, the policy-based ANASTACIA framework aims to cope with research challenges related with interoperability and scalability IoT security management. That is, the policy-based approach aims to deal with the heterogeneity and scalability by defining different level of abstractions, models and translation plugins. In this way, the scalability is also benefited since the policy-based approach with a high-level of abstraction makes easier to manage a large amount of devices. The policy conflict detection allows the framework to deal with several conflict types, and finally the policy for orchestration considers policy chaining by priority or dependencies to cover an orchestration plan.

Currently, the project is validating the related components and features by experimenting on IoT/SDN/NFV Proof of Concepts for different security capabilities like authentication, authorization and accounting (AAA), filtering, IoT management, IoT honeynet and channel protection as it can be seen in the research outcomes.

Regarding the research outcomes and associated publications, [2] provides a first PoC performance evaluation focused on a sensor isolation through different SDN controllers as well as a traditional firewall approach. [3] shows the potential of the policy-based framework focused on a AAA scenario. The paper entitled "Virtual IoT HoneyNets to mitigate cyberattacks in SDN/NFV-enabled IoT networks" shows the dynamic deployments of IoT-honeynet networks on demand by replicating real IoT environments by instantiating the ANASTACIA IoT-honeynet policy model. It also provides performance for different kind of IoT devices and topologies. In [1], the authors present the architecture focusing on the reaction performance of the policy-based framework.

2.3.2 Investigation on Innovative Cyber-threats

The CNR team involved in ANASTACIA has multi-year experience in the cyber-security field, concerning both the development of innovative cyber-attacks and intrusion detection algorithms. By exploiting the knowledge of the team, in the ANASTACIA context, deep work has been accomplished in the cyber-security context. Such work led to the identification of two innovative threats, related to the IoT and Slow DoS Attacks contexts. The novelty of such threats is demonstrated by their acceptance from the research world [4, 5]. In the following, based our description on the published works just mentioned and on the descriptions reported in the project deliverables, the introduced new attacks are briefly described (how they work and how it is possible to protect from them).

2.3.2.1 IoT 0-day attack

Being exchanged information extremely sensitive, due to the nature of IoT devices and networks, security of IoT systems is a topic to be investigated in deep. The work behind the proposed attack goes in this direction, by investigating the domotic IoT context and exploiting its components, to identify weaknesses that attackers may exploit.

The proposed attack is part of the ZigBee security context. ZigBee is a wireless standard introduced by the ZigBee Alliance in 2004 and based on the IEEE 802.15.4 standard, used in the Wireless Personal Area Networks (WPAN) context [6]. In particular, we identified a particular vulnerability affecting AT Commands capabilities implemented in IoT sensor networks. Our work focuses on the exploitation of such weakness on XBee devices, supporting remote AT commands, exploited to disconnect an end-device from the ZigBee network and make it join a different (malicious) network and hence forward potentially sensitive data to third malicious parties. Given the nature of IoT end-devices, often associated with a critical data and operations, it may be obvious how a Remote AT Command attack represents a serious threat for the entire infrastructure. Early evaluation of the effects of the proposed attack on a real network led to validate the success of the proposed threat [4]. Obtained results prove the efficacy of the proposed attack.

Moreover, since just a single packet is sent to the victim by the attacker to reconfigure it, the proposed attack should be considered as dangerous as scalable. Particularly, the time required to send such packet is minimal, so in case of multiple targeted sensors, the attack success is guaranteed.

By adopting an external level protection approach [4], the protection system is directly employed on the nodes, since agents implemented on the IoT devices are responsible for monitoring the device status and verifying that all the parameters are correct. In case the device is affected by a remote AT reconfiguration command attack, such alert information is forwarded to the IoT coordinator, and the device is designed to mitigate the attack (by autonomously reconfiguring itself, as previously described). Since not all the devices may embed a detection and mitigation system, the IoT coordinator is supposed to also monitor devices status periodically to identify disconnections, hence report them to the other ANASTACIA modules.

2.3.2.2 Slow DoS attacks

Among all the methodologies used to successfully execute malicious cyber-operations, denial of service attacks (DoS) are executed with the aim of exhaust victim's resources, compromising the targeted systems' availability, thus affecting availability and reliability for legitimate users. These threats are particularly dangerous, since they can cause significant disruption on network-based systems [7]. The term Slow DoS Attack (SDA), coined by the CNR research group involved in the project, concerns a DoS attack which makes use of low-bandwidth rate to accomplish its purpose. An SDA often acts at the application layer of the Internet protocol stack because the characteristics of this layer are easier to exploit to successfully attack a victim even by sending it few bytes of malicious requests [8]. Moreover, under an SDA, an ON-OFF behaviour may be adopted by the attacker [9], which comprises a succession of consecutive periods composed of an interval of inactivity (called off-time), followed by an interval of activity (called on-time).

The innovative attack proposed is called SlowComm, sending a large amount of slow (and endless) requests to the server, saturating the available connections at the application layer on the server inducing it to wait for the (never sent) completion of the requests. As an example, we refer to the HTTP protocol, where the characters sequence \r\n\r\n represent the end of the request: SlowComm never sends such characters, hence forcing the server to an endless wait. Additionally, during a SlowComm the request payload is sent abnormally slowly. Similar behaviour could be adopted for other protocols as well (SMTP, FTP, etc.). As a consequence, by applying this behaviour to a large amount of connections with the victim, a DoS may be reached. In particular, SlowComm works by creating a set of predefined connections with the victim host. For each connection, a specific payload message is sent (the payload is typically endless), one character at time (one single character per packet), by making use of the Wait Timeout [9] to delay the sending. In this way, once the connection is established with the server (at the transport layer), a single character is sent (hence, establishing/seizing the connection at the application layer, hence, with the listening daemon). At this point, the Wait Timeout is triggered, to delay the sending of the remaining payload, and to prevent server-side connection closures. During our work we proved how the attack may successfully lead a DoS to different popular TCP based services [4], hence proving that the attack is particularly dangerous.

To protect from SlowComm and Slow DoS Attacks in general, it is important to consider the following fact: *it is trivial to detect and mitigate a single attacking host, while it is extremely difficult to identify a distributed attack*. This fact derives from the fact that IP address filtering may be applied to detect and mitigate a SlowComm attack (see, for instance, our tests on mod-security [4]), while in case of a distributed attack this concept may not be adopted with ease. Moreover, from the stealth perspective, the proposed attack is particularly difficult to detect while it is active, since log files on the server are often updated only when a complete request is received or a connection is closed: being our requests typically endless, during the attack log files do not contain any trace of attack. Therefore, different approaches should be adopted, for instance based on statistic [10], machine learning [6, 11, 12], or spectral analysis [13]. A possible approach to adopt combines the algorithm proposed in [10] and the methodology proposed in [14] to detect running SlowComm attacks. Early version of the algorithms has been tested in laboratory, while testing on relevant environments has not been accomplished to date. Concerning the ANASTACIA platform, further work on the topic will be focused on evaluating a possible implementation of such approach, aimed to provide protection from Slow DoS Attacks by embedding innovative anomaly-based intrusion detection algorithms in a relevant environment and providing additional capabilities to the ANASTACIA framework, in the context of cyber-security applied to counter last generation threats.

2.3.3 Trusted Security Orchestration in SDN/NFV-enabled IOT Scenarios

In the ANASTACIA architecture, the security orchestrator oversees orchestrating the security enablers according to the defined security policies. The later would be generated either by the end-user or received from the monitoring and reaction plane. The security orchestration plane, through its components security orchestrator, security resource planning and policy interpreter, is able to coordinate the policies and security enables to cover the security configuration needed for different communications happen in the network. The security orchestration plane takes into account the policies requirements and the available resources in the underlying infrastructure to mitigate the different attacks while reducing the expected mitigation cost and without affecting the QoS requirements of different verticals. The resources in the underlying infrastructure refer to the available amount of resources in terms of CPU, RAM, and storage in different cloud providers, as well as the bandwidth communication between these network clouds.

Figure 2.3 depicts the main architecture of the security orchestration and enforcement plane suggested in ANASTACIA. Using SDN network, the IoT domain is connected to the cloud domain, whereby different IoT services are running. The user accesses the IoT devices, first, through the cloud domain, then the SDN enabled network and the IoT router. In fact, in ANASTACIA, the communication between a user and an IoT device happens through a chain of virtual network functions (VNFs) named service function chaining (SFC). The latter consists of three parts:

  (i) The ingress point, which is the first VNF in the SFC. The user initially attaches to the ingress point;
 (ii) The intermediate VNFs;
(iii) The egress point, which is the last VNF in the SFC. The egress point should be connected to the IoT controller. As depicted in Figure 2.3, the order of the communications between the VNFs is defined according to the different SDN rules enforced thanks to the SDN controller. The nature and the size of the SFC would be defined according to the nature of the user (a normal or a suspicious).

Figure 2.4 depicts the different steps of the orchestration and enforcement plane suggested in ANASTACIA. The attack is detected thanks to the Mitigation Action Service (MAS) component. The later sends a mitigation request (MSPL file) to the security orchestrator (Figure 2.4, Step 3). To mitigate the attacks, the security orchestrator interacts with three main actors, which are (Figure 2.4):

IoT controller: It provides IoT command and control at high-level of abstraction in independent way of the underlying technologies. That is, it is able to carry out the IoT management requests through

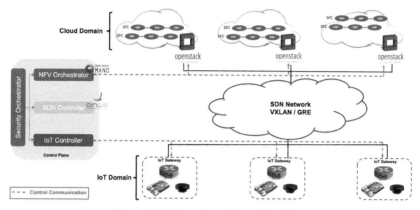

Figure 2.3 Security orchestration plane.

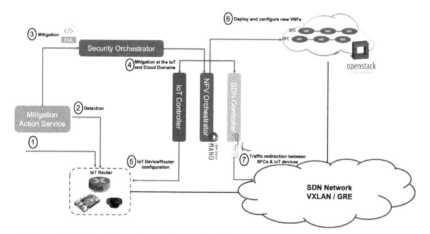

Figure 2.4 Security orchestration and enforcement in case of a reactive scenario.

different IoT constrain protocols like CoAP or MQTT. It also maintains a registry of relevant information of the deployed IoT devices like the IoT device properties and available operations. Since it knows the IoT devices status, it could be able to perform an effective communication to avoid the IoT network saturation when it is required a high-scale command and control operation. In "Security Management Architecture for NFV/SDN-aware IoT Systems" (Under review) can be found an example and performance of IoT management as part of a building management system. To mitigate different attacks, the security orchestrator interacts with the IoT controller to mitigate the attacks at the level of the IoT domain and prevent the propagation of the attack to other networks (Figure 2.4: 4). The IoT controller enforce different security rules at the IoT router (data plane) to mitigate the attack (Figure 2.4: 5).

NFV orchestrator: In ANASTACIA, to ensure efficient management of SFC, we have integrated SDN controller (ONOS) with the used Virtual Infrastructure Manager (VIM), in our case OpenStack. The integration of SDN with the VIM enable the smooth communication between different VNFs that form the same SFC. After receiving the MSPL message from the MAS, the security orchestrator identifies the right mitigation plane should be implemented. If the mitigation plan requires the instantiation of new VNFs, the security orchestrator instructs the NFV orchestrator to instantiate and configure the required VNFs. To instantiate the required VNFs, the NFV orchestrator interacts with the VIM (Figure 2.4: 6). Also, the security orchestrator interacts with the policy interpreter to translate the received MSPL to the low configuration (LSPL) needed for different VNFs. After the successful instantiation of a security VNF, the security orchestrator configures that VNF with the received LSPL (Figure 2.4: 6).

In ANASTACIA, we have also developed different virtual security enablers that should be instantiated to mitigate the different attacks. For instance, we have developed a new VNF firewall based on SDN-enabled switch and OpenFlow. OVS-Firewall is a newly developed solution that relies on OpenFlow protocol to create a sophisticated firewalling system. We have also proposed and developed a new security VNF, named virtual IoT-honeynet, that allows to replicate a real IoT environment in a virtual one by simulating the IoT devices with their real deployed firmware, as well as the physical location. The IoT-honeynet can be represented by an IoT-honeynet security policy, and the final config- uration can be deployed transparently on demand with the support of the SDN network. "Virtual IoT HoneyNets to mitigate cyberattacks in SDN/NFV-enabled IoT networks" (Under review) shows the potential and performance of this approach.

SDN controller: This component helps in rerouting the traffic between the VNFs in different SFCs. As depicted in Figure 2.4, when the mitigation action service notifies the orchestrator about an attack, the SFC would be updated by adding/inserting new security VNFs in the SFCs. The security orchestrator should push the adequate SDN rules to reroute the traffic between different VNFs in the SFC and the IoT domain (Figure 2.4: 7). Also, according to the different situations, the security orchestrator can choose the SDN as security enabler. In this case, it can be the attack mitigated by pushing exploring the strength of the SDN technology. If so, the security orchestrator can instruct the SDN controller to push some SDN rules to prevent, allow or limit the communication on specified protocols and ports between different communication peers (Figure 2.4: 7).

By relying in the aforementioned orchestration properties and features, as well as the SDN and IoT controllers, the ANASTACIA framework aims to cope with the research challenges related with Orchestration of SDN/NFV-based security solutions for IoT environments and currently several experiments have been carried out in different security areas.

For instance, several experiments have been carried out regarding **virtual IoT-honeynets**. This kind of VNF allows to replicate a real IoT environment in a virtual one by simulating the IoT devices with their real deployed firmware, as well as the physical location. The IoT-honeynet can be represented by an IoThoneynet security policy, and the final configuration can be deployed transparently on demand with the support of the SDN network. "Virtual IoT HoneyNets to mitigate cyberattacks in SDN/NFV-enabled IoT networks" (paper under review) shows the potential and performance of this approach.

Furthermore, the security orchestration of ANASTACIA enables continuous and dynamic management of Authentication, Authorization, Accounting (AAA) as well as Channel Protection virtual security functions in IoT networks enabled with SDN/NFV controllers. Our scientific paper [1] shows how a virtual AAA is deployed as VNF dynamically at the edge, to enable scalable device's bootstrapping and managing the access control of IoT devices to the network. Besides, our solution allows distributing dynamically the necessary crypto-keys for IoT M2M communications and deploy virtual Channel-protection proxies as VNFs, with the aim of establishing secure tunnels (e.g. through DTLs) among IoT devices and services, according to the contextual decisions inferred by the cognitive framework. The solution was implemented and evaluated, demonstrating its feasibility to manage dynamically AAA and channel protection in SDN/NFV-enabled IoT scenarios.

A telco cloud environment may consist of multiple VNFs that can be shipped and provided, in the form virtual machine (VM) images, from different vendors. These VNF images will contain highly sensitive data that should not be manipulated by unauthorized users. Moreover, the manipulation of these VNF images by unauthorized users can be a threat that can affect the whole system setup. In ANASTACIA, we have designed and developed different tools to prevent the manipulation of different VNF images should run on top of different network clouds. In ANASTACIA, we have devised efficient methods that verify the integrity of physical machines before using them and also the integrity of virtual machine and virtual network function images before launching them [15–17]. For this purpose, different technologies have been investigated, such as i) Trusted Platform Module (TPM); ii) Linux Volume Management (LVM); iii) Linux Unified Key Setup (LUKS). For instance, in [16], we have provided a trusted cloud platform that consists of the following components:

- TPM module that is used to store passwords, cryptographic keys, certificates, and other sensitive information. TPM contains platform configuration registers (PCRs) which can be used to store cryptographic hash measurements of the system's critical components. There are in total 24 platform configuration registers (PCRs) in most TPM modules starting from 0 till 23.
- Trusted boot module, which is an open source tool, uses Intel's trusted execution technology (TXT) to perform the measured boot of the system. Trusted boot process starts when trust boot is launched as an executable and measures all the binaries of the system components (i.e., firmware code, BIOS, OS kernel and hypervisor code). Trust boot then writes these hash measurements in TPM's secure storage.
- Remote attestation service, which is the process of verifying the boot time integrity of the remote hosts. It is a software mechanism integrated with TPM, to securely attest the trust state of the remote hosts. It uses boot time measurements of the system components such as BIOS, OS, and hypervisor, and stores the known good configuration of the host machine in its white list database. It then queries the remote host's TPM module to fetch its current PCR measurements. After receiving the current PCR values, it compares them against its white list values to derive the final trust state of the remote host.
- OpenStack Resource Selection Filters component that should be integrated with the nova scheduler. In OpenStack, when a VNF is launched, the nova-scheduler filters pass through each host and select the number of hosts that satisfy the given criteria. Each filter passes the list of selected hosts to the proceeding filter. When the last filter

is processed, OpenStack's default filter scheduler performs a weighing mechanism. It assigns weight to each of the selected hosts depending on the RAM, CPU and any other custom criteria to select a host which is most suitable to launch the VM instance.

2.3.4 Dynamic Orchestration of Resources Planning in Security-oriented SDN and NFV Synergies

Network operators are facing different type of attacks that introduce new set of challenges to detect and to defend from the attack. However, the hardware appliances for defence or detection are neither flexible nor elastic and they are expensive. To extend the NFV MANO framework, ANASTACIA incorporates a set of intelligent and dynamic security policies that can be updated seamlessly to constantly reflect security concerns in the VNF placement through the resource planning module while still ensuring acceptable QoE. Moreover, we have defined and implement synergies between SDN controllers and NFV MANO for the purpose of coordinating security to have an effective impact by defining adequate SDN rules or the adequate virtual security appliances (VNF) to be enforced through the Security Enabler Provider module. In the following section the resource planning and the security enabler provider modules will be defined.

2.3.4.1 Resource planning module

During the first phase of ANASTACIA, we have done two main works. The first one focused on the selection of best service (Virtual Network Function (VNF)), called "The security enablers selection", among the list of enablers selected previously by the selected Security Enabler Provider, to cope with a security attack, and a second work focus "Mobile Edge Computing Resources Optimization". In fact, one of our two main use cases focuses on Mobile Edge Computing, as an example, to secure protection of a company perimeter, based on several buildings with different usage situated in different areas using distributed resource as MEC; an emerging technology that aims at pushing applications and content close to the users (e.g. at base stations, access points, and aggregation networks), reduces the latency, improves the quality of experience, and ensures highly efficient network operation and service delivery.

During the second phase of the project, we aim to extend the resource planning module to include a dynamic Service Function Chain (SFC) requests placement that aim to reduce the routing overhead in case of an attack happen as an example. In fact, it is challenging to allocate multiple SFC requests on an NFV Infrastructure, especially in a cost-driven objective. VNFs have to be chained in a specific order. Moreover, depending on their type and isolation considerations, VNFs can be potentially shared among several SFCs. Finally, VNFs must not be placed far from the shortest path to avoid increasing SFC delay and network usage.

2.3.4.2 The security enablers selection

The aim of the model is to select the best service (Virtual Network Function (VNF)) among the list of enablers selected previously by the selected Security Enabler Provider, to cope with a security attack and that minimize the maximum load nodes (CPU, RAM, bandwidth) of the topology, provided by the system model. Indeed, the system information will provide relevant data about the whole infrastructure, server capacity (CPU, RAM, etc.), and VNF flavours (CPU, RAM, etc.). On the other hand, the Security Enablers information will provide the data regarding the available Security Enablers capable to enforce specific capabilities. The goal of the model is minimizing the maximum load nodes to improve provider cost revenue (provider energy efficiency goal). For more details please refer to the Anastacia deliverable D3.3.

2.3.4.3 Mobile edge computing resources optimization

Mobile edge computing (MEC) is an emerging technology that aims at push-ing applications and content close to the users (e.g. at base stations, access points, aggregation networks) to reduce latency, improve quality of experi-ence, and ensure highly efficient network operation and service delivery. It principally relies on virtualization-enabled MEC servers with limited capac-ity at the edge of the network. One key issue is to dimension such systems in terms of server size, server number and server operation area to meet MEC goals. In this work, we have proposed a graph-based algorithm that, taking into account a maximum MEC server capacity, provides a partition of MEC clusters, which consolidates as many communications as possible at the edge. We evaluate our proposal and show that, despite the spatio-temporal dynamics of the traffic; our algorithm provides well-balanced MEC areas that serve a large part of the communications.

This work has been published in a Sigcomm [18] workshop and extended for a TNSM journal [19].

2.3.4.4 Security enabler provider

The Security Enabler Provider is a component of the Security Orchestration Plane, as defined in the Anastacia architecture. This component is able to identify the security enablers which can provide specific security capabilities, to meet the security policies requirements. Moreover, when the Security Resource Planning, a sub-component of the security orchestrator, defined before, selects the security enabler, the Security Enabler Provider is also responsible for providing the corresponding plugin.

The Security Enabler Provider primarily interacts with the Policy Interpreter. Specifically, two different interactions have been contemplated:

- The first one will provide to the Policy Interpreter a list of security enabler candidates from the main identified capabilities.
- The second one will provide to the Policy Interpreter the specific Security Enabler Plugin to perform the policy translation. This policy translation process was defined in Anastacia D3.1 [20], and also published in journal paper [2].

The first role is implemented as a piece of software that from the specific capabilities given as an input it will provide the more accurate enablers. The second role is also implemented as piece of software capable to translate MSPL policies into specific configuration/tasks rules according to a concrete security enabler. For more details please refer to the Anastacia deliverable D3.3 [21].

2.3.5 Security Monitoring to Threat Detection in SDN/NFV-enabled IOT Deployments

Security threat levels change dynamically as the attackers discover new breaches and try to exploit them. To cope with this challenge, the ANASTACIA project relies on SDN and NFV techniques to embed the developed security products and provide a dynamic way to deploy them when needed. In this way, the ANASTACIA project delivers a set of scientific and technological innovations, grouped in two principal key innovation areas.

2.3.5.1 Security monitoring and reaction infrastructure

Saedgi et al. identify the principal challenges when securing IoT-based Cyber Physical Systems, highlighting as one of the principal challenges the development of a "*a holistic cybersecurity framework covering all abstraction layers of heterogeneous IoT systems and across platform boundaries*" [22]. The ANASTACIA project fulfils this challenge by proposing a state-of-the-art security infrastructure composed by three principal modules:

- *Monitoring Agents*: These are the components in charge of extracting the security data from the monitored network. The ANASTACIA framework has been designed flexible enough to support both physical and virtual monitoring agents, as well as to extract data from data networks (both IP and IoT networks) and from analogue CPS devices. This make the ANASTACIA framework a multilevel security platform, and therefore suitable for physical sensor networks, emulated environments and hybrid networks. In this direction, the ANASTACIA partners have worked in the implementation of monitoring agents adapted for 6LowPan and ZigBee IoT networks, as well as the development of agents capable of extracting temperature information from analogue sources. These agents have been tested using the case studies of the project, aiming to be applied in wider scenarios for its final validation. Following this path, the project partners are extending even further these monitoring agents with virtualization characteristics. By means of using NFV and SDN technologies on the monitoring agents, it will be possible to deploy and (re)configure them on demand, allowing to deploy new agents on the network as a reaction to ongoing attacks. In this sense, the ANASTACIA partners are also extending the security policy language (MSPL) to correctly specify such type of countermeasures, allowing the deployment of new monitoring agents on the network in a complete autonomous manner.

- *Monitoring Module*: This component contains the logic of the detection of security incidents. The heterogeneous monitoring agents (IoT networks and analogue agents) use a shared communication channel to publish the extracted security data. This information is then analysed by the incident detectors (for well-known attacks) and behaviour analysis modules (for zero-days attacks), emitting verdicts about the detected incidents. As stated in [22], detecting zero-days attacks does not ensure a high security level, since well-known attacks are still used by malicious users to gain control of the systems. ANASTACIA does not only provide both types of analysis (well-known attacks and behaviour analysis) but it will also use all this information to provide a deeper analysis and found correlations between already-known attacks and they behavioural analysis result, detecting hidden relationships between events coming from

different sources. The ANASTACIA partners are developing such correlation engines to enhance both security analyses and provide enriched
information to the reaction module.

- ***Reaction Module***: Using the information provided by the monitoring
  module (namely incidents verdicts and behavioural analysis results), the
  reaction module has the responsibility of determining the best mitigation
  plan for the detected incidents. The ANASTACIA framework provides a
  simple yet powerful design for this component, which uses not only the
  incidents verdicts provided by the monitoring module, but also system
  model and the capabilities deployed in the network. All this information
  is enhanced with a risk analysis to determine the best set of countermeasures to cope with the ongoing attack. Further information about how
  this analysis is performed can be found in the following sections.

2.3.5.2 Novel products for IoT- and cloud-based SDN/NFV systems

The security infrastructure described above represents one of the principal
outcomes of the project, however the partners are also working on a concrete
implementation of this design. To implement this monitoring infrastructure,
the partners have developed a set of technologies that fulfil the functionalities of the ANASTACIA infrastructure, generating a set of novel products
ready to be deployed on IoT- and cloudbased systems. For example, partner
Montimage has developed a 6LowPan network sniffer in coordination with
the MMT tool to detect anomalies in IoT networks. UTRC (in collaboration
with OdinS) has developed analogue temperature agents and a machine
learning-based behavioural analysis for data sensors, allowing them to detect
zero-days attacks on temperature sensor networks. ATOS has extended its
XL-SIEM tool to perform the risk analysis when computing the reaction
and the inclusion of the system model when computing the countermeasures
to be taken. Despite the development of such products is not finished yet,
the partners have managed to integrate PoC version of such technologies
on a shared platform, allowing to perform initial tests and validation of the
technologies. Moreover, it is envisaged to further extend this tools with a
correlation engine, aiming to reveal hidden relationships between security
events coming from different sources (monitoring agents) and, therefore,
raising the awareness level of the whole security platform.

   To further extend the offer of products, the ANASTACIA partners are
preparing the solutions to be NFV- and SDN-ready, by means of adapting

the solutions (especially network agents) to work as single, self-contained NFV modules. In this sense, the ANASTACIA outcomes will have the potential to be deployed in virtualized environments, be dynamically deployed as a reaction to an ongoing attack and, capable of being reconfigured if required. In this scenario, the ANASTACIA platform will have the ability to momentarily harden the security of the portions of the network are under attack, by means of deploying new agents, load new security rules on the monitoring agents/module, analyse new protocols or reconfigure the existing instances. All these actions are to be maintained until the security level has returned to normal values or the network administrator has intervened to solve the security breach.

All these novel products will have a high impact on the security market, opening business possibilities in the IoT-based CPS area.

Despite the ambition of the project is high, the ANASTACIA partners have already established the bases of the further innovations. The ANASTACIA partners will continue its efforts to fully integrate the security innovations with the SDN and NFV technologies, as well as developing a correlation engine for security events. This direction aims to provide the market with a highly-dynamic security solution, capable of not only detecting current cyber threats, but also capable of reacting against them and also deploy new security instances to adapt to the always-evolving security levels of IoT networks.

2.3.6 Cyber Threats Automated and Cognitive Reaction and Mitigation Components

The monitoring information and the incident detected are evaluated for automatic mitigation. Security policies are used to determine the security enablers supported by the IoT infrastructure. This is also used to know the mitigations that the IoT infrastructure supports. Obviously, not all mitigations work with all possible threats, and not all mitigations have the same cost. Cost is not considered here just in terms of economic impact, but also in terms of time to mitigate, computational resources required or complexity of the mitigation. ANASTACIA automatically analyses these factors and, along with the incidents detected, evaluates and decides on the most convenient mitigation in each case. To this end several data are considered in the analysis:

- severity of the incidents, which is received by the correlation engine at the monitoring module and takes into account the type of incident and the duration of the incident among others,

- importance of the assets affected, which depends on the criticality of the IoT devices affected, their location or the importance of the data they manage,
- the cost of the mitigation, obtained either from the orchestrator in charge of enforce the available security enablers, or from the system admin in case specific expert knowledge is required.

The global risk of the incident is obtained from (1) and (2), which is used together with (3) to decide on the most convenient mitigation. A decision support service (DSS) is used to compute that information, providing with a score for each mitigation, which represents the suitability of the mitigation for the ongoing incident. The mitigation with the higher suitability score represents the most suitable mitigation, which is passed to the orchestrator for its enforcement. To this end a Mitigation Action Service (MAS) is used to translate the output of the DSS to a format that is understandable by the orchestrator. The MAS is then in charge of generating the reaction in the MSPL format. This language was selected since its XML-based structure allows specifying the type of base capability to deploy (e.g. filtering, monitoring), and the configurations of such action (e.g. involved IPs, port numbers, number of agents to deploy). The MSPL format also allows the MAS to directly send the mitigation plan to the Security Orchestrator, which will use it to deploy the computed plan.

In order to generate the MSPL file, the MAS analyses the response of the DSS by performing the following processes: (1) it identifies the countermeasure computed by the DSS; (2) it identifies the network capabilities able to execute the countermeasure; (3) it retrieves the information of the capabilities from the System Model Analysis module; (4) it builds the MSPL file to express the countermeasure, specifying the capability to use and the configurations of that capability used to apply the countermeasure.

Every incident handled by the reaction (including risk evaluation, decision support activities), the information associated to it (such as type of incident or IoT devices affected) and all the indicators that characterize the incident (such as severity, importance of assets affected, global risk of the incident or suitability of the mitigation) are passed to the Dynamic Security and Privacy Seal to update the seal status.

Currently we are developing the quantitative model that supports the assessment of incidents and mitigations for deciding on the most convenient reaction based on incident severity, criticality of the assets affected, possible mitigations and cost of mitigating them.

2.3.7 Behaviour Analysis, Anomaly Detection and Automated Testing for the Detection of Known and Unknown Vulnerabilities in both Physical and Virtual Environments

Our behavioural framework automatically identifies cyber-security attacks in a given IoT environment. It uses system design and operational data to discover dependencies between cyber systems and operations of HVAC in a cyber-physical domain. We predict potential security consequences of interacting operations among subsystems and generate threat alarms. Specifically, our behavioural engine is empowering ANASTACIA's use case scenario using the "best" practices to implement security in terms of (1) adding network security (in forms of IDS/IPS), and (2) using threat intelligence to detect evasions or hidden attacks. Our developed platform can detect:

- Known attacks such as DDoS and MiTM attacks,
- IoT zero-days attacks and slow DoS attacks that might pass undetected by normal IDS/IPS [9].

Our framework developed a monitoring component that is composed of messaging wrappers, Constraint Programming (CP) models and buffered sensor data from IoT networks. Primarily, CP model is the core component of our behavioural analysis engine. First the information is gathered and analysed for learning a CP model and then it is deployed to identify any intrusion. Moreover, CP model built on continuous stream of data (i.e. time-series) where the time interval between successive updates could vary from milliseconds to minutes. CP model consists of network of relations between building sensor data. Using this CP model, we aggregate the different types of sensor data to truly model the normal behaviour of the system that is being supervised. This model is built for monitoring at system level, but it does not prevent from including in the model information about network performance if that is exposed to it. For an example, CPU consumption of a device can be included along its actual sensor data. The variety of data that we can aggregate allows the model to be as generic or as specific as the end-user required it to be. Since the model is built on relations, we can leverage from the fact that what data effects what other data type (features).

We developed an approach to learn a CP-based decision model consisting of a set of relations to detect misbehaviour of the system. More specifically, the idea is to learn a set of relations which together when satisfied defines the normal behaviour of the system. After learning important relations, the approach discards un-important relations, and consequently creates a model with best possible relations and features of sensor nodes. In each iteration, the relation between the sensor features and all other network features further verified. Also, we identify the sensors are involved in breaking the relation and what are the set of relations are broken Following this fashion, the model is further tuned. The developed 'Monitoring' component enables continuous and integrated monitoring of multivariate signals, event logs, heartbeat signals, status reports, operational information, etc., emanating from various devices in multitude of building operational subsystems. This monitoring component also evaluates the security situation against known policies, models, threat signatures to detect abnormalities and outliers, e.g. high data download, external database or port accesses during an emergency. Such situations will be analysed by the 'Reaction' component which will evaluate the severity of the situation. Isolation and predictive mechanisms are activated to ensure that the rest of the building operations system continues as normal. Policies and rules are activated, updated and enforced by the 'Security Enforcement' component, e.g. a building emergency will lock-down the non-essential database accesses, and escalation of the emergency to the city fire brigade should be performed by any of the authorized personnel. To this end, our behavioural engine's innovation is summarized as the following key points:

- Learning constraint programming model for capturing the normal behaviour of a given cyberphysical system
- CP-model provides explanation when a potential anomaly is detected by reporting which constraints fails to satisfy the model
- User-defined constraints can be easily integrated with the constraints learn from the data
- The developed behaviour engine can handle multiple attacks of different types.

2.3.8 Secured and Authenticated Dynamic Seal System as a Service

Several projects have tried to address the need to enable trustable ICT deployments. The solutions they have developed are generally focused either on enhancing trust on security or on privacy, but not both. This situation can be counterproductive if considered in the context of the obligations emerging from the recently adopted European General Data Protection Regulation (GDPR) (which considers both security and privacy controls as fundamental to the protection of personal data).

Moreover, existing solutions are usually based on two separate models:

- Either ISO standard-based certification of products and information management systems respecting ISO 17065 or ISO 17021-1 and relaying on human audit and assessment;
- Or purely system-based monitoring of security, such as anti-virus applications or intrusion detection system (IDS), which are often designed independently from any standard.

The ever-evolving normative framework for security and personal data protection calls for a holistic approach which considers technical insights alongside human and organizational controls. An organization that seeks to comply with the regulatory frameworks will finally rely on the professional advice from information security professionals (spearheaded by a Chief Information Security Officer -CISO-) and legal professionals (usually taking the token as Data Protection Officers -DPO-), which might have difficulties understanding the complex outputs of the technological enablers used to introduce the necessary controls to the systems they oversee and integrating these with the legal and managerial feedback necessary to transparently and accurately demonstrate due diligence has been carried out.

In response to this situation, ANASTACIA's Dynamic Privacy and Security Seal (DSPS) will seek to inform the end-user (DPO/CISO) on the most relevant privacy and security issues while supporting certification and compliance activities. To this end, the DSPS will:

- Introduce a privacy-by-design and by default compliant architecture, services and graphical user interface (GUI) that seek to combine the certainty and trustworthiness of conventional certification schemes with real-time certification surveillance capabilities through the real time dynamic monitoring (provided by ANASTACIA) of the certified system.
- Compile alerts and threats from ANASTACIA, compatible monitoring solutions (using the STIX 2 standard) and the end-user (CISO/DPO) and showcase them through a unified GUI, displaying IoT/CPS privacy and security information while providing decision support capabilities, and data visualization (considering accessibility/ease of use require-ments).
- Empower the end-user by enabling the client's Data Protection Offi-cer (DPO) and Chief Information Security Officer (CISO) to provide feedback to the raised alerts directly through the GUI and to enhance the information obtained from the monitoring system with technical, legal, annd organizational documentation. This data will be stored in a -privacy-by-design- distributed storage solution (powered by Shamir

Secret Sharing Scheme), which will be associated with the DSPS blockchain-based seal ledger (Hyperledger Fabric), to ensure the data is non-repudiable, immutable, and easily verifiable in direct relation to the events showcased by the DSPS both by the end-user (for internal audit and compliance purposes) and associated certification bodies (to determine the validity of relevant certifications).

The Dynamic Security and Privacy Seal (DSPS) aims to provide a holistic solution to privacy and security monitoring, addressing both the organizational and technical requirements enshrined by the GDPR through the implementation of a layered process by which: 1) an initial examination by an auditor or expert determines the baseline status of the system with regards to privacy and security of both the product or system that is to be monitored, and the organizational policies and mechanisms that surround its implementation to ensure compliance with the most relevant ISO standards (particularly if linked to a certification) and regulations; 2) ANASTACIA provides constant monitoring and reaction capabilities which are then used to update the DSPS; 3) the end-user provides feedback on the effectivity of the mitigation activities and uses the DSPS enablers to enhance transparency and accountability in the monitored system.

The resulting tool will provide the end-user with a broad perspective over the state of the monitored system which will consistently track and unify the organizational/human elements considered by personal data protection regulations with the technical insights provided by ANASTACIA's monitoring and reaction services. Once implemented, this process will not only provide advanced trust-enhancing information functionalities to ANASTACIA users, but will also serve as a surveillance solution for audit/certification/legal compliance purposes. It will generate a non-repudiable historic track of system variations and potential threats (technical and organizational) to the sealed system while enhancing the contextual information available to the client, auditors or regulatory authorities.

Current work [23] has been focused towards developing the DSPS architecture as defined by ANASTACIA Deliverable 5.1; deploying and integrating the monitoring service and associated enablers; and refining the GUI elements that will inform the end-user and enable them to provide the required feedback. Upcoming research will seek out ways to simplify complex privacy and security information, so as to address the varying technical and legal knowledge of the potential end-users. Furthermore, research on integration with additional information sources (particularly through the

STIX2 format) and privacy-management tools (such as the CNIL DPIA software) will be performed to further enhance the functionalities available through the DSPS GUI.

2.4 Conclusion

This book chapter has summarized the main key innovations being devised, implemented and validated in the scope of Anastacia research project to meet the cybersecurity challenge in heterogenous IoT scenarios. Namely it has presented eight key innovations: 1) Holistic policy-based security management and orchestration in IoT, 2) Investigation on innovative cyber-threats, 3) Trusted Security orchestration in SDN/NFV-enabled IoT scenarios, 4) Dynamic orchestration of resources planning in Security-oriented SDN and NFV synergies, 5) Security monitoring to threat detection in SDN/NFV-enabled IoT deployments, 6) Cyber threats automated and cognitive reaction and mitigation components, 7) Behaviour analysis, anomaly detection and automated testing for the detection of known and unknown vulnerabilities in both physical and virtual environments, 8) Secured and Authenticated Dynamic Seal System as a Service.

These main key innovations are currently being realized and evaluated successfully in MEC and Smart-building scenarios. In this sense, important research outcomes have been already obtained and published in high impact journals, which demonstrate the feasibility and performance of ANASTACIA cybersecurity framework to dynamically handling and counter evolving kind of cyberattacks in SDN/NFV-enabled IoT deployments.

Acknowledgements

This work has been supported by the following research projects:

- Advanced Networked Agents for Security and Trust Assessment in CPS/IoT Architectures (ANASTACIA), funded by the European Commission (Horizon 2020, call DS-01-2016) Grant Agreement Number 731558.

The authors declare that there is no conflict of interest regarding the publication of this document.

References

[1] Alejandro Molina Zarca, Jorge Bernal Bernabe, Ruben Trapero, Diego Rivera, Jesus Villalobos y, Antonio Skarmeta, Stefano Bianchi, Anastasios Zafeiropoulos and Panagiotis Gouvas "Security Management Architecture for NFV/SDN-aware IoT Systems", IEEE IoT Journal, 2019.

[2] Molina Zarca, A.; Bernal Bernabe, J.; Farris, I.; Khettab, Y.; Taleb, T.; Skarmeta, A. Enhancing IoT 719 security through network softwarization and virtual security appliances. International Journal of Network 720 Management, 28, e2038, https://onlinelibrary.wiley.com/doi/pdf/10.1002/nem.2038.e2038, 721 doi:10.1002/nem.2038.

[3] Molina Zarca, Alejandro and Garcia-Carrillo, Dan and Bernal Bernabe, Jorge and Ortiz, Jordi and Marin-Perez, Rafael and Skarmeta, Antonio, Enabling Virtual AAA Management in SDN-Based IoT Networks, Sensors, 19, 2019, 2, 295, http://www.mdpi.com/1424-8220/19/2/295, 1424-8220, 10.3390/s19020295

[4] Cambiaso, E., Papaleo, G., and Aiello, M. (2017). Slowcomm: Design, development and performance evaluation of a new slow DoS attack. Journal of Information Security and Applications, 35, 23–31.

[5] Vaccari, I., Cambiaso, E., and Aiello, M. (2017). Remotely Exploiting AT Command Attacks on ZigBee Networks. Security and Communication Networks, 2017.

[6] Katkar, V., Zinjade, A., Dalvi, S., Bafna, T., and Mahajan, R. (2015, February). Detection of DoS/DDoS Attack against HTTP Servers Using Naive Bayesian. In Computing Communication Control and Automation (ICCUBEA), 2015 International Conference on (pp. 280–285). IEEE.

[7] Beitollahi, H., and Deconinck, G. (2011). A dependable architecture to mitigate distributed denial of service attacks on network-based control systems. International Journal of Critical Infrastructure Protection, 4(3–4), 107–123.

[8] Cambiaso, E., Papaleo, G., and Aiello, M. (2012, October). Taxonomy of slow DoS attacks to web applications. In International Conference on Security in Computer Networks and Distributed Systems (pp. 195–204). Springer, Berlin, Heidelberg.

[9] Cambiaso, E., Papaleo, G., Chiola, G., and Aiello, M. (2013). Slow DoS attacks: definition and categorisation. International Journal of Trust Management in Computing and Communications, 1(3–4), 300–319.

[10] Aiello, M., Cambiaso, E., Scaglione, S., and Papaleo, G. (2013, July). A similarity based approach for application DoS attacks detection. In Computers and Communications (ISCC), 2013 IEEE Symposium on (pp. 000430–000435). IEEE.

[11] Duravkin, I. V., Carlsson, A., and Loktionova, A. S. (2014). Method of slow-attack detection. Системи обробки інформації, (8), 102–106.

[12] Singh, K. J., and De, T. (2015). An approach of DDOS attack detection using classifiers. In Emerging Research in Computing, Information, Communication and Applications (pp. 429–437). Springer, New Delhi.

[13] Brynielsson, J., and Sharma, R. (2015, August). Detectability of low-rate HTTP server DoS attacks using spectral analysis. In Advances in Social Networks Analysis and Mining (ASONAM), 2015 IEEE/ACM International Conference on (pp. 954–961). IEEE.

[14] Cambiaso, E., Papaleo, G., Chiola, G., and Aiello, M. (2016). A Network Traffic Representation Model for Detecting Application Layer Attacks. International Journal of Computing and Digital Systems, 5(01).

[15] S. Lal, A. Kalliola, I. Oliver, K. Ahola, and T. Taleb, "Securing VNF Communication in NFVI," in Proc. IEEE CSCN'17, Helsinki, Finland, Sep. 2017.

[16] S. Lal, I. Oliver, S. Ravidas, T. Taleb, "Assuring Virtual Network Function Image Integrity and Host Sealing in Telco Cloud," in Proc. IEEE ICC 2017, Paris, France, May 2017.

[17] S. Lal, T. Taleb, and A. Dutta, "NFV: Security Threats and Best Practices," in IEEE Communications Magazine., Vol. 55, No. 8, May 2017, pp. 211–217.

[18] M. Bouet, V. Conana, Geo-partitioning of MEC resources, ACM MECOMM '17, August 21, 2017, Los Angeles, CA, USA.

[19] M. Bouet, V. Conana, Mobile Edge Computing Resources Optimization: A Geo-Clustering Approach, IEEE Transactions on Network and Service Management, Vol. 15, No. 2, June 2018.

[20] AM Zarca, JB Bernabe, AS, K Yacine, B Dallal, S Bianchi "Initial Security Enforcement Manager Report". 2018. H2020 Anastacia EU project deliverable D3.1.

[21] D Belabed, M Bouet, D Rivera, P Sobonski, A Molina Zarca, "Initial Security Enforcement Enablers Report" Anastacia EU project deliverable D3.3.

[22] Sadeghi, Ahmad-Reza, Christian Wachsmann, and Michael Waidner. "Security and privacy challenges in industrial internet of things." Design

Automation Conference (DAC), 2015 52nd ACM/EDAC/IEEE. IEEE, 2015.

[23] Quesada Rodriguez, Adrian; Bajic, Bojana; Crettaz, Cédric; Filipponi, Matteo; Pacheco Huamani, Ana María; Perlini, Adriano, Kim, Eunah; Loup, Vincent; Ziegler, Sébastien. "Dynamic Security and Privacy Seal Monitoring Service". 2018. H2020 Anastacia EU project deliverable 5.2.

[24] Quesada Rodriguez, Adrian; Bajic, Bojana; Crettaz, Cédric; Menon, Mythili; Pacheco Huamani, Ana María; Kim, Eunah; Loup, Vincent; Ziegler, Sébastien. "Dynamic Security and Privacy Seal Model Analysis". 2018. H2020 Anastacia EU project deliverable 5.1.

[25] C. M. Ramya, M. Shanmugaraj, and R. Prabakaran, "Study on ZigBee technology," in Proceedings of the 3rd International Conference on Electronics Computer Technology (ICECT '11), pp. 297–301, IEEE, Kanyakumari, India, April 2011.

[26] J. Yick, B. Mukherjee, and D. Ghosal, "Wireless sensor network survey," Comput. networks, vol. 52, no. 12, pp. 2292–2330, 2008.

[27] Ziegler S. et al. (2019) Privacy and Security Threats on the Internet of Things. In: Ziegler S. (eds) Internet of Things Security and Data Protection. Internet of Things (Technology, Communications and Computing). Springer, Cham, DOI: 10.1007/978-3-030-04984-3_2

3

Statistical Analysis and Economic Models for Enhancing Cyber-security in SAINT

Edgardo Montes de Oca[1], John M. A. Bothos[2] and Stefan Schiffner[3]

[1]Montimage Eurl, 39 rue Bobillot, Paris, France
[2]National Center for Scientific Research "Demokritos", Patr. Gregoriou E. & 27 Neapoleos Str, Athens, Greece
[3]University of Luxemburg
E-mail: edgardo.montesdeoca@montimage.com; jbothos@iit.demokritos.gr; Stefan.schiffner@uni.lu

SAINT analyses and identifies incentives to improve levels of collaboration between cooperative and regulatory approaches to information sharing. Analysis of the ecosystems of cyber-criminal activity, associated markets and revenues drive the development of a framework of business models appropriate for the fighting of cyber-crime. The role of regulatory approaches as a cost benefit in cyber-crime reduction is explored within a concept of greater collaboration to gain optimal attrition of cyber-criminal activities. Experimental economics aid SAINT in designing new methodologies for the development of an ongoing and searchable public database of cyber-security indicators and open source intelligence. Comparative analysis of cyber-crime victims and stakeholders within a framework of qualitative social science methodologies deliver valuable evidences and advance knowledge on privacy issues and deep web practices. Equally, comparative analysis of the failures of current cyber-security solutions underpins a model for greater effectiveness and improved cost-benefits. SAINT advances the metrics of cyber-crime through the construct of a framework of a new empirical science that challenges traditional approaches and fuses evidence-based practices with more established disciplines. Innovative models, algorithms and automated framework for metrics benefit decision-makers, regulators, law enforcement, at national and organisational levels providing improved cost-benefit analysis and estimation of tangible and intangible costs for optimal risk and investment incentives.

3.1 Introduction

The SAINT project[1] examines the problem of failures in cyber-security using a multidisciplinary approach that goes beyond the purely technical viewpoint. Building upon the research and outcomes from preceding projects, it combines the insights gained to progress further analysis into economic, behavioural, societal and institutional views in pursuit of new methodologies that improve the cost-effectiveness of cyber-security.

SAINT analyses and identified incentives to improve levels of collaboration between cooperative and regulatory approaches to information sharing to enhance cyber-security and mitigate (a) the risk and (b) the impact from a cyber-attack, while providing, at the same time, solid economic evidence on the benefits from such improvement based on solid statistical analysis and economic models.

It is widely acknowledged that despite the sums spent annually on cyber-security, cyber-crime continues to flourish. No true or accurate picture of the situation is readily available and yet vast amounts of money continue to be employed in efforts to reduce levels of cyber-crime that do not appear to be working. There are now more than 3.6 billion Internet users[2] and 7.3 billion mobile-cellular subscriptions worldwide[3] in 2016 and rising. According to Microsoft's report [1] on "Cyberspace 2025: Today's Decisions, Tomorrow's Terrain", it is estimated that by 2025, more than 91% of people in developed countries and nearly 69% of those in emerging economies will be using the Internet, with the total number of Internet users estimated to be 4.7 billion. In this expanding cyber-space, it is estimated that at least 7% of URLs are malicious, 85% of the 200 billion emails processed per day are spam, 1.4 million browser agents are botnets, consisting 20% of mobile browser agents and measurable cyber-attacks rise up to 1 million plus every day. The

---

[1]SAINT (Systemic Analyser In Network Threats) is an H2020 project. See https://cordis.europa.eu/project/rcn/210229 and https://project-saint.eu for more information.

[2]www.internetworldstats.com (30 June 2016).

[3]www.itu.int

annual cost to the global economy from cyber-crime is €300 billion, with the average annualized cost of data breaches only, being €7.9 million. The global cyber-crime market represents €15 billion and up to €50 billion for security products and services [2]. Europol, in its 2015 report [3] "Exploring Tomorrow's Organized Crime" forecasts an expansion of cyber-crime, in the form of a project-basis, where cyber-criminals lend their knowledge, experience and expertise as part of a crime-as-a-service business model. The crime-as-a-service business model is facilitated by social networking, digital infrastructures and virtual currencies that allow cyber-criminals to exchange and use financial resources anonymously on a large scale.

The EU FP7 project CyberROAD[4] successfully delivered a research roadmap for cyber-crime and cyber–terrorism using in-depth analysis into technological, social, legal, ethical, political, and economic origins of the issues. A noted research outcome was the proposed innovative cyber-crime cost-benefit reduction methodology as delivered in the paper "2020 Cybercrime Economic Costs: No Measure No Solution", [2]. In furtherance of the insights already gained in the CyberROAD project, SAINT carries out an extensive analysis of the state-of-the-art using a range of comparative studies to deliver a framework of data-driven guidelines based on mathematical analysis of the relevant quantitative variables that decision makers require for accurate resource allocation. The construct of such a framework designed with experimental economics aligns and regulates the discipline to that of an empirical science and substantiates the case for greater collaboration in information sharing.

3.2 SAINT Objectives and Results

3.2.1 Main SAINT Objectives

SAINT project studies and improves the measurement approaches and methodologies by means of constructing a framework of a new empirical science, challenge traditional approaches and fuse evidence-based practices with more established disciplines for a lasting legacy. Through the construction this framework, it gives decision makers (public policy authorities, business leaders and individuals) data-driven guidelines based on scientific analysis of relevant quantitative and qualitative variables for their decisions about dedicating resources to deal with cyber-threat risks and cyber-criminals.

---

[4]https://www.cyberroad-project.eu

By employing various methodologies from different scientific fields, the main objectives of SAINT are to:

1. Establish a complete set of metrics for cyber-security economic analysis, cyber-security and cyber-crime market.
2. Develop new economic models for the reduction of cyber-crime as a cost-benefit operation.
3. Estimate and evaluate the associated benefits and costs of information sharing regarding cyber-attacks.
4. Define the limits of the minimum needed privacy and security level of internet applications, services and technologies.
5. Identify potential benefits and costs of investing in cyber-security industry as a provider of cyber-security services.
6. Develop a framework of automated analysis, for behavioural, social analysis, cyber-security risk and cost assessment.
7. Provide a set of recommendations to all relevant stakeholders including policy makers, regulators, law enforcement agencies, relevant market operators and insurance companies.

3.2.2 Main SAINT Results

The SAINT project examines the problem of failures in cyber-security using a multidisciplinary approach that combines economic, behavioural, societal and institutional approaches in pursuit of new methodologies that improve the cost-effectiveness of cyber-security. SAINT analyses and identifies incentives to improve levels of collaboration between cooperative and regulatory approaches to information sharing to enhance cyber-security and mitigate (a) the risk and (b) the impact from a cyber-attack, while providing, at the same time, solid economic evidence on the benefits from such improvement based on solid statistical analysis and economic models.

3.2.2.1 Metrics for cyber-security economic analysis, cyber-security and cyber-crime market

SAINT investigates and establishes accurate indicators and metrics for economic analysis, cyber-security and cyber-crime market, including the effects of regulatory analysis on the economics of cyber-security. It investigates all the open source intelligence methodologies and performs an analysis on the effect of those metrics in different scenarios and environments. The establishment of metrics for measuring privacy is also included in this effort.

With respect to the metrics and indicators (objective 1), SAINT analyses [4]: 19 open source cyber-security indicator datasets (including ENISA's top 15); two indicators of emerging cyber-threats; Blacklists, Blocklists and Whitelists; five insecurity indicators; nine security indicators; nine economic indicators; five open source intelligence methodologies for cyber-threats. It includes relevant examples, usage, statistics, and metrics for each of the above indicators.

SAINT also gathers and analyses [5] evidences from stakeholders, across multiple disciplines, with the objective to examine the problem of failures in cyber-security beyond a purely technical viewpoint and gain advanced knowledge on economics and cyber-security practices from the stakeholders, enabling the gaining of a better understanding of their needs and requirements and providing insights on cyber-security and product value for money. As a consequence of this analysis, FICORA (Finish regulator), is now proactively involved and cooperating in distributing a survey for Finland to gain supporting metrics in answer to an important question: why does Finland have one of the best quantitative track records in cyber-security, within the EU & G20[5]?

It was additionally observed as a result of a comparative analysis that the inclusion of the cost of time spent/lost by cyber-crime victims provided an important metric for ROI calculations. Results show:

- The cost of cyber-crime is estimated to be €30 billion (0.242% of EU's GDP).
- The cost in time lost or spent in 2017 due to cyber-crime amounts to an estimated €60 billion.
- Therefore, the actual total cost of cyber-crime to the EU in 2017 can be estimated to be €90 billion.

3.2.2.2 Economic models for the reduction of cyber-crime as a cost-benefit operation

Significant effort of SAINT is dedicated in the research and development of new economic models for cyber-security and cyber-crime. A rich econometric and mathematical theoretical framework is implemented for this purpose, and the final methodologies and models are validated in a controlled environment under the supervision of the Hellenic Police Cyber-Crime Unit.

In relation to objective 2, research focuses on the organisation's effective operational processes [6] to achieve efficiency in production by investigating their incentives in choosing input combinations that minimise cost and,

---

[5]http://www.intercomms.net/issue-30/dev-3.html

consequently, maximise profits. With the rapid evolution of Cloud Internet, organisations have an alternative solution to substitute highly qualified Information Technology working staffs that are paid high wage rates, which means excessive labour costs, with subcontracting of such Information Technology services to external providers like the newly emerged Managed Service Providers Networks. In this way, organisations avoid the excessive economic investment costs to set up and develop in-house Information Technology departments from scratch and find the means to hire or offer professional training to existing working staff, with the potential risk of economic losses resulting, in case of failures from such internally structured departments. Research in this field concerns the organisations' decisions to substitute production factors, purchased in the respective production factor markets, to minimise their production cost. It studies the dependence of the organisations' policies, concerning the outsourcing of certain Information Technology activities, by purchasing Cloud Internet computer services from automated platforms of Managed Service Provider Networks, on the price of Information Technology labour force that is the wage rates in the Information Technology sector. The empirical research performed in showed that organisations' price cross-elasticity demand for Cloud Internet computer services is significantly negative towards the wage rate in the Information Technology sector for specialised Information Technology labour force by -21.84% ($\pm6.38\%$). The evolution of Cloud Internet in our time has given organisations many alternatives, especially in the area of Information Technology services that can be purchased online, through the participation in relevant automated platform networks, operated and managed by external providers, in the form of Managed Service Provider Networks.

SAINT identifies current cyber-security failures and requirements to improve the situation at all levels of cyber-security defences and across a variety of sectors [7]. It determines what constitutes a cyber-security failure, or what inadvertently increases the risk of a cyber-attack, using quantitative and qualitative analysis, to identify what new practices are required to improve cyber-security, reduce wasteful information technology spending and improve return on investment.

SAINT also investigates how cyber-attacks materialise, focusing on what lies behind and contributes to the materialisation of these attacks [8]. This basically represents the emergence of a whole new economy consisting of a new and fast-growing body of vulnerability markets with stakeholders selling and buying vulnerabilities to gain financial gains or avoid financial losses, associated with immaterial assets, namely the vulnerabilities and their exploits. The goal is to identify and categorise the vulnerabilities and exploits markets along with the involved stakeholders and their roles, to provide guidelines for cost-effective cyber-security methodologies that can be applied as counter-measures for defence against malicious hackers. Vulnerability announcements can inflict severe monetary and other intangible costs on the company's value.

3.2.2.3 Benefits and costs of information sharing regarding cyber-attacks

SAINT provides guidelines for information sharing between all the agents, for mitigating inefficiencies in the cyber-security investment landscape and in the total economy in general. These guidelines are based on the joint evaluation of measurable quantitative economic and technical variables regarding the influence of cyber-security information sharing in the cost structure, the rate of investment, the effective allocation of resources and the overall profitability of each agent.

SAINT estimates and evaluates the associated benefits and costs of information sharing regarding cyber-attacks (objective 3) [6, 9]. For this, international cooperation activities have been studied [9], such as the ITU Global Cyber-security Agenda (GCA).

The GCA is a framework launched in 2007 for international cooperation. It is designed for cooperation and efficiency, encouraging collaboration with and between all relevant partners and building on existing initiatives to avoid duplicating efforts. Within GCA, ITU and the International Multilateral Partnership Against Cyber Threats (IMPACT) promote the deployment of solutions and services to address cyber-threats on a global scale. It is a global multi-stakeholder and public-private alliance against cyber-threats. EU addresses cyber-security through tool policies that affect the structures and capabilities of organisations while in parallel takes action by providing incentives to support and promote the development of co-operation in the area of cyber-security, for detecting cyber-incidents and responding to cyber-attacks effectively and appropriately.

The Directive on the Security of Network and Information Systems, the "NIS Directive", mentioned as the first EU-wide cyber-security law, is designed among others to foster better co-operation in reporting serious incidents and adopting effective risk management practices.

Regarding the promotion of cooperation in cyber-security domain, ENISA also serves as a focal point for information sharing and spread of knowledge in the cyber-security community, through the setting up of

Information Sharing and Analysis Centres. Their role is particularly important in creating the necessary trust for sharing information between all the different agents.

The subject of co-operation between organisations and how it influences their effective performance and allocation of their resources in terms of decreasing production cost and profitable exploitation of production inputs has been studied [6]. In this context co-operation between organisations is defined as information sharing between them. It proves empirically the importance of co-operation through information sharing in minimising production cost and achieving economic efficiency in the allocation of resources. The associated benefits of information sharing between organisations have been evaluated. In the long-run, using information sharing processes for improving the production process has an almost −13% (± 3.58%) decreasing effect on the real (deflated) long-run average production cost for the sample of our Eurozone countries, for the time period 2009–2012.

3.2.2.4 Privacy and security level of internet applications, services and technologies

SAINT analyses the dependence of detection of cyber-security incidents, on behavioural features of network traffic flow to interpret adequately the careless behaviour of internet users, regarding the proper application of cyber-security norms and rules. For this, SAINT implemented a correlation analysis on quantitative technical and measurable qualitative behavioural variables, concerning network traffic flow characteristics and cyber-security behaviour characteristics.

Regarding the limits of the minimum needed privacy and security level of internet applications, services and technologies (objective 4), [10] devised models and mechanisms for measuring privacy and for user privacy protection mechanisms. Several formal frameworks of privacy notions, with differing assumptions, are proposed that study the relations between Anonymous Communication Networks and respective provided privacy.

Based on this work, in [18] SAINT proposes how these different frameworks can be unified by constructing a generalized indistinguishability game similar to the games used to define semantic security in cryptographic protocols [23].

Along with effective defences against website fingerprinting, such as continuous data flow, package padding and traffic morphing, adaptive padding between data packets with generic web traffic and clustering of webpages into similarity groups. Beyond this, SAINT investigates:

- Approaches for protecting publicly available databases like secure computation of elementary database queries, locally random reductions of sets to databases, zero Knowledge interactive (and non-interactive) proofs, data oblivious data transfers in private information retrieval.
- Privacy preserving credentials and authentication mechanisms like password-based authentication, cryptographic certificates, attribute-based credentials, electronic certificates and electronic Identities.
- Database content anonymisation concepts and techniques like k-anonymity, i-diversity, t-closeness, bloom filters, differential privacy.

3.2.2.5 Benefits and costs of investing in cyber-security

SAINT provides guidelines and frameworks for maximising efficiency in cyber-security services. Part of the effort is dedicated in the development of alternative ways and methods to get valuable information in measurable quantitative form of metrics and then to analyse it to highlight guidelines for competitiveness and profitability in the cyber-security industry. SAINT also determines the value of the underground and cyber-crime market within a wider investigation of information security markets including.

In relation to objective 5, SAINT proposes new models and new paradigms in cyber-security with a special focus on the incentives of the different stakeholders in the ecosystem of cyber-criminality. It was first necessary to identify the existing business models that cyber-criminals use, and to describe the different national strategies of European countries that have been put in place to fight against cyber-crime. From this, new models are proposed that provide innovative ways that help reduce cyber-crime by targeting the right incentives of both cyber-criminals and cyber-security practitioners. These models are compared among each other and their practical relevance is evaluated [11]. Some of the results obtained concern: the analysis of existing cyber-criminal business models; the analysis of national, European and international cyber-security policies and strategies and the draft of 8 innovative models to fight against cyber-crime, including: the certification and labelling services model; the insurance model; the wage model; the collaborative model; the education model; the crowdsourcing model; the bug-bounty model; the artificial intelligence model.

In relation to objective 5, SAINT demonstrates [8] that behind the materialisation of the cyber-attacks there is a new and fast-growing body of vulnerability markets with stakeholders selling and buying vulnerabilities for financial gains or to avoid financial loss. This implies that a whole new economy is rapidly evolving based on immaterial assets, the vulnerabilities and their exploits. Over the last years, ransomware attackers demanded payment in cryptocurrencies, with the Bitcoin[6] being among the most popular ones. Bitcoin offers anonymity in terms of involved parties and the amount of the transaction and their use for illicit purposes has become popular.

The "Execute Code"-related vulnerabilities are prevalent among all other vulnerabilities, which implies that software vendors (mainly OS developers) fail to take appropriate measures during the design and implementation stages. Most of the discovered vulnerabilities (over 50%) are severe, with a severity score at least six. This, in turn, may imply severe financial or other intangible (e.g. trust, fame) costs on affected companies. No software product or system is immune to vulnerabilities, which demonstrates that vulnerability discoverers could virtually target any vendor, operating system, or software product as long as it is either, (or both), a challenging or profitable target.

Vulnerability announcements can inflict severe monetary and other intangible costs (e.g. loss of trust and tarnished fame) on the affected company, measured by system downtime, operation disruption, loss of credibility and customers, higher assurance costs, etc.

Vulnerability announcements can lead to a negative and significant change in a software vendor's market value. According to the conducted quantitative analysis, an affected vendor can lose even 60% value in stock price when a related vulnerability is disclosed. Study has also showed that a software vendor loses more market share if the market is competitive or if the vendor is small. Moreover, as can be expected, the change in stock value is more negative if the vendor fails to provide the right patch at the time of disclosure of the vulnerability. In addition, according to the findings, key vulnerabilities have significantly more impact on the company's value.

Useful insights on the types of attacks per business sector have also been obtained [12]. Small businesses (with fewer than 250 employees) are those most targeted by cyber-attacks, making up as much as 43% of all the cyber-attacks on companies (in 2015). Large enterprises (with over 2,500 employees) accounted for 35% of all cyber-attacks, while medium-sized businesses (with between 251 and 2,500 employees) made up the remaining 22%. It is interesting to note that these results are diametrically opposed to those from 2011 where large businesses accounted for the majority (50%) of all cyber-attacks on companies, medium-sized businesses represented 32%, while small businesses accounted for 18%. Between 2011 and 2015, small businesses have been increasingly targeted by cyber-attacks. This trend can

---

[6]https://www.bitcoin.com/

be explained by the fact that, unlike big businesses that have the capacity to invest in proper expertise and technologies, smaller businesses may not always have the financial resources and staff to protect themselves from such threats. Consequently, cyber-attackers take advantage of smaller companies' digital vulnerability to steal confidential data and intellectual property, bring down the website, or organising phishing and spamming campaigns. Regarding the type of cyber-attacks on businesses, we have the following specificities:

- Spam: the size of a company has limited influence over its spam rate. Indeed, in 2016, the spam-rate varied between 52.6% and 54.2%, which shows that all kinds of companies are likely to be targeted, regardless of their size. Furthermore, all industry sectors receive similar quantities of spam.
- Phishing: although the overall phishing rates have declined over the past three years, companies are still targeted by these attacks. Medium-sized businesses experience the highest phishing rates. In 2016, the sector of agriculture, forestry, and fishing was the most affected by phishing, with one in 1,815 emails being classed as a phishing attempt.
- Data breaches: In 2016, the industry of services (particularly business services and health services) was the most affected by data breaches, representing 44.2% of all breaches. The sector of finance, insurance, and real estate was ranked second with 22.1%.

The private sector, particularly the cyber-security industry, plays an important role in combatting cyber-crime by providing individual users, businesses, and organisations with services and solutions to cyber-threats. In 2003, the global cyber-security market represented $2.5 billion, currently it amounts to $106 billion, and the sector will be worth $639 billion in 2023. These numbers underline the growing demand for cyber-security solutions and highlight the business opportunities in the sector.

In 2016, the commercial cyber-security vendors' market was dominated by the United States with a total of 827 vendors leading cyber-security research and products. Israel and the United Kingdom hold second and third place in the ranking with 228 and 76 vendors respectively.

While the cyber-security industry has potential for growth, in both the private and public sectors, it is still struggling to keep up with cyber-crime for three reasons:

- The variety of IoT devices: the increase in connected IoT devices increases the number of potential targets. Projections suggest that, by 2020, there will be tens of billions of connected digital devices in the EU alone.

- The multiplicity of data: an increase in connected IoT devices directly correlates with an increase in data that needs to be protected.
- The shortage of skilled workers in the cyber-security sector: in spite of the great employment opportunities and high number of open positions for IT specialists and cyber-security professionals, the cyber-security industry struggles with training them in time to keep up with growing demand. The solution to this problem may come from artificial intelligence and machine learning, which are currently being developed.

SAINT also performed a cost-benefit analysis of cyber-security solutions and products (objective 5). This is built on a cash flow analysis of cyber-security solutions, products and models. It relies on information from a market analysis established [12], on the revenue analysis of cyber-security services [13] and on the most relevant models identified. It also uses input from conducted surveys [14] and estimates the price of digital assets and the costs of intangible risks. In addition to the cash flow analysis, a sensitivity and risk analysis is implemented [15]. These recommendations serve as guidelines for various stakeholders, including cyber-security business providers. It builds on the cost-benefit implemented, as well as on the econometric analysis of cyber-security solutions, the market analysis, and the assessment of the innovative cyber-security models analysed [16].

3.2.2.6 Framework of automated analysis, for behavioural, social analysis, cyber-security risk and cost assessment

In the framework of automated analysis (objective 6), SAINT defines the different tools that constitute the Framework (Figure 3.1, [17]). This includes the cyber-security cost-benefit analysis tools and algorithms. Based on available metrics, indicators and parameters, the techniques allow the construction of models and the estimation of the price of digital assets and costs of intangible risks (e.g. reputation, non-critical service disruption). A toolset for automated analysis based on automatic information gathering and analysis tools that extract information from a variety of information sources on the Internet and the Deep Web has been designed and prototypes implemented. The tools include: Social Network Analyser and the Deep Web Crawler. The information sources include cyber-security related discussion forums, bug bounties, social network discussions and public vulnerability and data breach incident databases.

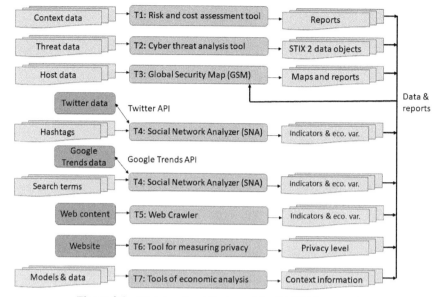

Figure 3.1 High-level architecture of the SAINT framework.

The developed Twitter Social Network Analyser (SNA) utilizes the social network, Twitter, to extract trends on the cyber-crime activity. To this end, a dictionary of #hashtags of interest is created. The SAINT SNA mines only publicly available tweets and accounts for the specific hashtags and extracts the related information.

The Google Trends SNA utilizes the popular Google Trends platform to extract trends that are related to cyber-crime activity. Google Trends is a public web facility of Google Inc. It is based on Google Search and shows how often a particular search term is entered with respect to the total searches in different regions of the world and in various languages.

Crawling and Scraping the Web and Deep Web can be categorized into two different large types, where each one includes a number of considerations and design decisions, depending to the target web sites that are searched (Web and Deep/Dark Web). The first type is Web Scraping of a website and the second one Crawling. The Tor network was found to be the ideal place for investigating cyber-criminal activity while browsing anonymously Deep Web sites to avoid of being hacked or traced. For the implementation of our scripts, we run Tor in the background to avoid being detected by users of the Deep and the Dark Web.

SAINT's Global Security Map (GSM)[7] gathers data on selected ENISA indicators using a variety of suitable open source feeds and presents the results visually on a global map. It is an interactive tool which enables visualization of the geographic distribution of the sources of cyber-crime and quantitative comparative metrics, with the aim to provide a simple and accurate method of displaying the global hotspots for the location and quantification of the top cyber-threat indicators: malware, phishing, spam, cyber-attacks, and other malicious activities. The unique combination of detailed data and simplified visualizations make the tool ideal for research and comparative analysis purposes by governments, law enforcement, CERTs, academia, Infosec, financial institutions and the public sector (also related to objective 7).

One more tool developed in the scope of SAINT project is Tool for measuring privacy in encrypted networks [18]. Resent research [19, 20] showed that user's privacy can be endangered even if he is using anonymization networks such as TOR [21] or JAP [22]. By means of an attack known as *website fingerprinting*, it is possible to identify which website a user is visiting and, thereby, to identify both two communicators and the content of the communication. However, different websites have different degrees of finger printability. Thereby, SAINT developed a tool which allows any user to estimate his vulnerability level to the website fingerprinting attack when visiting a website. Afterward, the user can decide if visiting this website costs the possible risks.

3.2.2.7 Recommendations to stakeholders

Reference model (Figure 3.2, [12]) illustrates the interactions between the different stakeholders involved in the cyber-crime and cyber-security ecosystem.

Related to objective 7, SAINT provides a set of recommendations to all relevant stakeholders (policy-makers, regulators, law enforcement agencies, relevant market operators and insurance companies) [16]. This builds on the input of various sources from different partners, including the stakeholder surveys that were conducted. An initial set of recommendations has been defined that includes:

- Adopting in-depth comparative analysis for the application of successful practices of individual countries, i.e. Finland (see Figure 3.3).

---

[7]https://3hz6pq.staging.cyberdefcon.com/

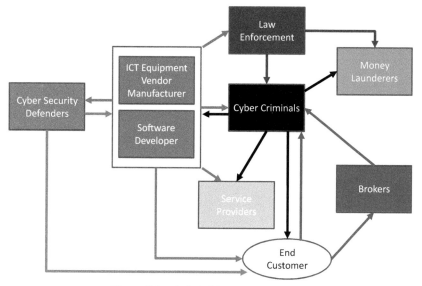

Figure 3.2 Stakeholder reference model.

- Improving the cost of cyber-crime metrics and econometrics for enhanced ROI calculations by the inclusion of the time spent or lost by cyber-crime victims.
- Improving the transparency of cyber-security matters within the workplace.
- Educating the workforce on the costs and risks to the workplace of cyber-practices.
- Furthering cyber-security training & education within the EU to alleviate the acknowledged lack of trained staff.
- Improving the complementarity among standards and best practices in cyber-security within the EU.
- Standardising the metrics to enable accurate comparative analysis between surveys/reports.

In Finland, FICORA has the role of a CERT that is a regulator but also acts to prevent and remediate cyber-security issues. The problem in other countries is that the regulators are only telecom regulators whereas in Finland FICORA is both a telecom and cyber-security regulator. Telecom operators are not really concerned about the security of customers. They just want to make sure that their services work, that the pricing brings profits and that the competition is regulated to their advantage. Most CERTs in Europe have a limited role that

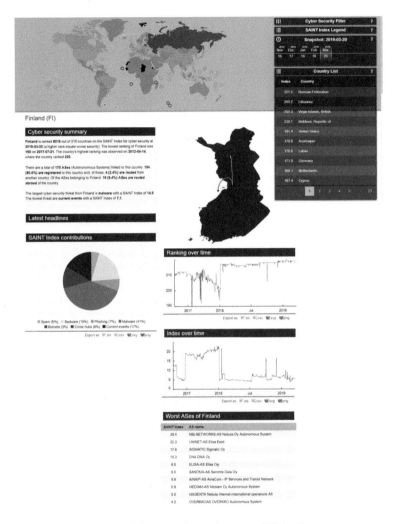

Figure 3.3   Global security map of Finland.

consists in reporting threats, building cyber-threat intelligence frameworks, and stimulating or developing cyber-threat solutions. When the safety and security of citizens is concerned we need entities that act and are proactive as is the case in the health and food sectors. FICORA is the best qualitatively and quantitatively. It bases its cyber-security activity on technologically efficient techniques, such as darknets or reverse network telescopes, but also it obtains results through sound organisation, clear objectives, close collaboration with all the stakeholders, and has the budget to do it.

Another aspect that should be emphasised is the legal one. The U.S. Anti-Bot Code of Conduct (ABC) for Internet Services Providers (ISPs) has resulted in an almost immediate reduction of botnets in the US. Operators started taking down botnets and collaborating to do so. What can be derived from this experience is that when a law is passed that identifies responsibilities and penalties, companies and individuals are incentivised. Telecom operators will start taking down botnets and fighting cyber-criminality only when it becomes financially interesting for them. Unfortunately, there is yet no law in Europe that is equivalent to ABC. The examples of collaborative actions since 2014 [9] show progress but the need remains to obtain a more systematic approach for fighting cyber-crime that is better and more globally organized. This can only be achieved with effective laws, regulations, incentivisation and cooperation at the national and international levels.

3.3 Conclusion

The SAINT project has worked on and advanced in the comprehension of the stakes involved in the cyber-security domain. It has analysed the risks and cost of security threats by compiling a complete set of metrics for the analysis of cyber-security economics, cyber-security risks, and the cyber-crime market. New economic models and algorithms have been developed to find optimised cost-benefit solutions for reducing cyber-crime.

The deep analysis of the benefits obtained from cyber-attacks information sharing (in particular, cooperative and regulatory approaches), positive impact of investments in cyber-security by industry, and the risks and costs of security breaches have resulted in a set of recommendations valuable for all relevant stakeholders (e.g. policy makers, regulators, law enforcement agencies, industry). The studies and surveys conducted have also allowed to better understanding the limitations and needs involved when finding the equilibrium between privacy and security of internet-based applications, services and technologies.

SAINT has also developed a framework that facilitates the automated analysis for behavioural, social, cyber-security risk and cost assessment. Research gaps have been addressed that can help policy makers make more informed decision on where economic investments should be directed to return the best possible outcomes. The different tools that constitute the SAINT Framework target improving the automation of certain analysis tasks and present the results in an integrated way, at least partially. The resulting system serves as a proof of concept that will show the usefulness of the integration of data from different sources and tools. In the future, the Framework will be extended and include a tighter integration so that researchers can process different types of security intelligence information and obtain results in a methodical way.

The main challenge identified by SAINT is to find the best approaches to:

- Coordinate cyber-security related issues and actions (i.e., related to legislative, regulatory, law enforcing and cooperative) between different organisations and countries;
- Measure the effectiveness of the actions;
- Achieve long-term impact to improve the security of ICT users;
- Implement and enforce laws and regulations in a virtualised and often conflicting international context;
- Make security an integral part of ICT design;
- Reverse the tendency that makes economic incentives better for criminals that those who need to protect their systems;
- Achieve consensus between stakeholders and countries;
- Improve education related to cyber-security;
- Find a good balance between security and privacy.

Having analysed different regulations and practices our conclusion is that we need to attack the cyber-threat problem from all fronts at the same time, in other words we need to:

- Improve the laws and regulations and make them more comprehensive;
- Coordinate better the regulatory processes and incentivize cooperation;
- Make cyber-security and privacy protection an obligation of service providers (including operators) to their customers;
- Greatly improve the awareness of the individuals to the risks;
- Change the economics to reduce the benefits of cyber-criminal activities and improve the perceived benefits of cyber-security measures. This includes reforming the international finance system to eliminate, or at least greatly reduce, the money laundering possibilities (e.g., tax havens, bitcoins).

Many of the challenges are addressed in the case of Finland, except maybe for the challenges related to the privacy concerns and the economics and financial aspects. Collaborative actions need to be done in a more systematic, global and organised way for fighting cyber-crime. This can only be achieved with effective laws, regulations, incentivisation and cooperation at the national and international levels. Currently, cyber-crime is more incentivized and even cooperates better than organisations that fight it. This situation needs to be reversed and obtaining profits by cyber-criminals should be made much more complicated.

Acknowledgements

This work is performed within the SAINT Project (Systemic Analyser in Network Threats) with the support from the H2020 Programme of the European Commission, under Grant Agreement No. 740829. It has been carried out by the partners involved in the project:

- National Center for Scientific Research "Demokritos", Integrated Systems Laboratory, Greece
- Computer Technology Institute and Press DIOFANTUS, Greece
- University of Luxembourg, Luxemburg
- Center for Security Studies – KEMEA, Greece
- Mandat International, Switzerland
- Archimede Solutions SARL, Switzerland
- Stichting CyberDefcon Netherlands Foundation, Netherlands
- Montimage EURL, France
- Incites Consulting SARL, Luxemburg

Thus, we would like to thank the different contributors from these organizations:

Eirini Papadopoulou, Konstantinos Georgios Thanos, Andreas Zalonis, Constantinos Rizogiannis, Antonis Danelakis, Ioannis Neokosmidis, Theodoros Rokkas, Dimitrios Xydias, Jart Armin, Bryn Thompson, Jane Ginn, Pantelis Tzamalis, Vasileios Vlachos, Yannis Stamatiou, Marharyta Aleksandrova, Latif Ladid, Dimitris Kavallieros, George Kokkinis, Cesar Andres, Christopher Hemmens, Anna Brékine, Sebastien Ziegler, Olivia Doell, Gabriela Znamenackova, Gabriela Hrasko.

We would also like to thank FICORA (Finland) for their help.

References

[1] Burt, Kleiner, Nicholas, Sullivan, "Cyberspace 2025 Today's Decisions, Tomorrow's Terrain, Navigating the Future of Cyber-security Policy", Microsoft Corporation, June 2014.
[2] Armin, Thompson, Kijewski, Ariu, Giacinto, Roli, "2020 Cybercrime Economic Costs: No measure No solution", 10th International

Conference on Availability, Reliability and Security, Toulouse, August 2015.

[3] European Police Office, "Exploring Tomorrow's Organised Crime", 2015 available at: https://www.europol.europa.eu

[4] Jart Armin, Bryn Thompson et al., "Final report on Cyber-security Indicators & Open Source Intelligence Methodologies", SAINT D2.1 Deliverable. Not yet available.

[5] Jart Armin, Bryn Thompson et al., "Final Report on the Comparative Analysis of Cyber-Crime Victims", SAINT D2.3 Deliverable. Not yet available.

[6] John M.A. Bothos et al., "Cyber-security Empirical Stochastic Econometric Modelling of Information Sharing and Behavioural Attitude", SAINT D3.1 Deliverable available at: https://project-saint.eu/deliverables

[7] Bryn Thompson, Jart Armin et al., "Final Analysis on Cyber-security Failures and Requirements", SAINT D3.3 Confidential Deliverable.

[8] Yannis Stamatiou et al., "Analysis of Legal and Illegal Vulnerability Markets and Specification of the Data Acquisition Mechanisms", SAINT D3.5 Deliverable available at: https://project-saint. eu/deliverables

[9] Edgardo Montes de Oca, Cesar Andres et al., "Comparative Analysis of Incentivised Cooperative and Regulatory Processes in Cyber-security", SAINT D2.5 Deliverable available at: https://project-saint.eu/deliverables

[10] Stefan Schiffner, Marharyta Aleksandrova et al., "Metrics for Measuring and Assessing Privacy of Network Communication", SAINT D2.6 Deliverable available at: https://project-saint.eu/deliverables

[11] Olivia Döll, Gabriela Hrasko et al., "Business Modelling Report", SAINT D4.3 Deliverable. Not yet available.

[12] Christopher Hemmens, Anna Brékine et al., "Stakeholder and Ecosystem Market Analysis", SAINT D4.1 Deliverable available at: https://project-saint.eu/deliverables

[13] John M.A. Bothos, Eirini Papadopoulou, Konstantinos Georgios Thanos, "Cyber-security and Cyber-crime Market & Revenue Analysis", SAINT D4.2 Deliverable. Not yet available.

[14] Bryn Thompson et al., "Stakeholder and Consumer Requirements Survey Report", SAINT D6.2 Deliverable available at: https://project-saint.eu/sites/deliverables

[15] Theodoros Rokkas, Ioannis Neokosmidis, Dimitris Xydias et al., "Report on Cost-Benefit Analysis of Cyber-security Solutions, Products and Models", SAINT D4.4 Deliverable. Not yet available.

[16] Archimede Solutions et al., "Recommendations on Investment, Risk Management and Cyber-Security Insurance", SAINT D4.5 Deliverable. Not yet available.

[17] Edgardo Montes de Oca, Cesar Andres et al., "Requirements Specification & Architectural Design of The SAINT Tool Framework", SAINT D5.1 Deliverable available at: https://project-saint.eu/deliverables

[18] Stefan Schiffner, Marharyta Aleksandrova et al., "Semi-automated Traffic Analysis of Encrypted Network Traffic", SAINT D5.2. Not yet available.

[19] A. Panchenko, L. Niessen, A. Zinnen, and T. Engel, "Website fingerprinting in onion routing based anonymization networks," in Proceedings of ACM WPES. Chicago, IL, USA: ACM Press, pp. 103–114, October 2011.

[20] A. Panchenko, A. Mitseva, M. Henze, F. Lanze, K.Wehrle, and T. Engel, "Analysis of fingerprinting techniques for tor hidden services," in Proceedings of the Workshop on Privacy in the Electronic Society, pp. 165–175, ACM, 2017.

[21] R. Dingledine, N. Mathewson, and P. Syverson, "Tor: The second generation onion router," in Proceedings of USENIX Security, San Diego, CA, USA: USENIX Association, 18 p, 2004.

[22] O. Berthold, H. Federrath, and S. Kopsell, "Web mixes: A system for anonymous and unobservable internet access," in Proceedings of Designing Privacy Enhancing Technologies: Workshop on Design Issues in Anonymity and Unobservability, pp. 115–129, July 2000.

[23] C. Kuhn, M. Beck, S Schiffner, T. Strufe, and E. Jorswieck "Privacy framework for anonymous communication", 20 pages, in print (poPETS, 2019).

4

The FORTIKA Accelerated Edge Solution for Automating SMEs Security

Evangelos K. Markakis[1], Yannis Nikoloudakis[1], Evangelos Pallis[1], Ales Černivec[2], Panayotis Fouliras[3], Ioannis Mavridis[3], Georgios Sakellariou[3], Stavros Salonikias[3], Nikolaos Tsinganos[3], Anargyros Sideris[4], Nikolaos Zotos[4], Anastasios Drosou[5], Konstantinos M. Giannoutakis[5] and Dimitrios Tzovaras[5]

[1]Department of Informatics Engineering, Technological Educational Institute of Crete, Greece
[2]XLAB d.o.o., Slovenia
[3]Department of Applied Informatics, University of Macedonia, Greece
[4]Future Intelligence LTD, United Kingdom
[5]Information Technologies Institute, Centre for Research & Technology Hellas, Greece
E-mail: Markakis@pasiphae.teicrete.gr; Nikoloudakis@pasiphae.teicrete.gr; Pallis@pasiphae.teicrete.gr; ales.cernivec@xlab.si; pfoul@uom.edu.gr; mavridis@uom.edu.gr; geosakel@uom.edu.gr; salonikias@uom.edu.gr; tsinik@uom.edu.gr; Sideris@f-in.co.uk; Zotos@f-in.co.uk; drosou@iti.gr; kgiannou@iti.gr; tzovaras@iti.gr

4.1 Introduction

Although the recent trend for the term "cyber-attack" is restricted for incidents causing physical damage, it has been traditionally used to describe a broader range of attempts to make unauthorized use of an asset related to computer information systems, computer networks, or even personal computing devices. As such, a cyber-attack aims to steal, alter a targets' system/data, or even destroy targets by gaining access into a targeted system. In this respect, a whole new industry has been shaped around the need for protection against cyber-attacks, i.e. the "cyber-security" domain,

77

which primarily deals with the protection of systems (incl. HW/SW & data) connected to the internet against cyber-attacks and should not be necessarily mixed with the domain of Information Technology (IT) Security (see Figure 4.1) that mainly refers to the protection of information. Cyber-security, on the other hand, is the ability to protect or defend the use of cyberspace from cyber-attacks by securing "things", vulnerable through ICT.

The first cyber-attack was recorded in 1989, in the form of a computer worm (i.e. malware), while their number has significantly grown in the following years (see Figure 4.2). Equal growth has been noted in the level

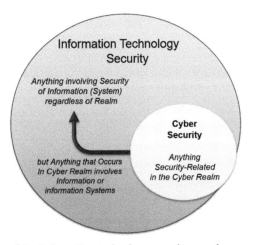

Figure 4.1   Information technology security vs cyber-security.

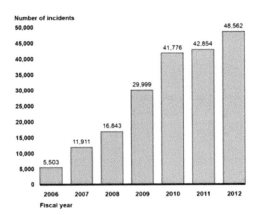

Figure 4.2   Incidents reported to US-CERT, Fiscal Years 2006–2014.
(*Source: GAO Analysis data of US-CERT*).

of both the threat they pose and the sophisticated manner with which they are launched and/or acting. Specifically, cyber-security threats have evolved from standalone threats that could affect single targets, to more complicated scenarios, where threats could be self-replicated, mutated and expanded to other devices and/or networks via the internet. Finally, the evolution of the exploitation manner of modern cyber-attacks is also extremely interesting. For instance, traditional ways for (i) harming infrastructures through DDoS attacks, (ii) misusing them through malwares, and (iii) mitigating them through identity spoofing are nowadays considered outdated and new emerging threats and attack scenarios are emerging, which aims at disusing sensitive soft assets through ransomware that directly lead their endangerment and their potential loss.

Unavoidably, cyber-security becomes, of great importance due to its increasing reliance on computer systems. Recently, in the era of the Internet of Things (IoT)[1], a large number of connected devices, located at the edge of the Internet hierarchy, generate massive volumes of data at high velocities. This turns centralized data models non-sustainable, since it is infeasible to collect all the data to remote data centres and expect the results to be sent back to the edge, with low latency.

Based on the constantly increasing dependency of the global economy on inter-connected digitization (i.e. world-web-web, smart-grids, IoT nets, direct communication links between platforms, etc.), it is the integrity and the availability of the prompt & uninterrupted interconnectivity that attracts great focus and investment from major players in the market. Similarly, the trend towards IoT and digital innovation, forms a flourishing business landscape for SMEs. However, this is put at stake due to the uncertain, cumbersome and most importantly costly nature of holistic cyber-security solutions. Specifically, although tailored solutions capable of providing the appropriate cyber-security levels for big companies appear, they can hardly be adapted to other environments and thus, lack in scalability, which makes them unsuitable for smaller enterprises.

The implementation of a complete and reliable edge computing security framework seems to be a promising alternative to protect an IoT environment and the overall network of an SME. In order to fulfil the IoT requirements, modern trends dictate that more resources (incl. computation, storage &

---

[1]IoT: The network of physical devices with connectivity (i.e. connect, collect & exchange data). The term was first introduced, when the amount of connected devices outnumbered the humans connected to the internet.

networking) must be located closer to users and the IoT devices, at the edge of the networks, where data is generated (i.e. *"edge computing"*), so as to (i) reduce data traffic especially in Internet backbone, (ii) provide in-situ data intelligence, (iii) reduce latency[2] and (iv) improve the response speed.

This way, cyber-security solutions will become more, in terms of both applicability and adaptability per use-case. Toward this direction, monolithic approaches are not enough; in-situ analysis based on usage Behaviour Analytics and Security Information & Event Management (SIEM) systems, customized at the for certain edge, seem able to offer a plausible and affordable solution, if offered as a modular product of adequate granularity in terms of offered services, so as to form an attractive product, easily customizable to the needs of the each customer.

Toward this direction, edge solutions introduce 5 major challenges[3] that require attention, namely (i) the massive numbers of vulnerable IoT devices, (ii) the NFV-SDN integrated edge cloud platform, (iii) the privacy & security[4] of the data, (iv) the interaction between edge & IoT devices and (v) the Trust & Trustworthiness.

This chapter presents an analysis on the cyber-threats landscape within generic ICT environments and its impact on SMEs, it also covers the different standardization and certification schemas that would help SMEs to support a cyber-security strategy and takes into consideration, standardization and best practices for the FORTIKA ecosystem and deployment. Additionally, the modular, edge-based cyber-security solution of the FORTIKA concept[5] is promoted within the current article. The resources required from a potential SME customer are efficiently managed, while a dedicated marketplace is a repository that can extend the basic version product with affordable functionalities tailored to the needs of each SME. On top of the latter, one can

---

[2]Given the complexity of cyber-security tasks and the latency imposed by the network distance between the client and the cloud infrastructure, one can deduce that cloud computing architecture, is by-design unsuitable for time-sensitive applications. The advancements in Edge Computing [1–3] allow for the efficient deployment and delivery of minimum-latency services.

[3]J. Pan, Z. Yang, "Cybersecurity Challenges and Opportunities in the New "Edge Computing + IoT" World", Association for Computing Machinery, 2018, doi: 10.1145/3180465.3180470

[4]Business data can be either sensitive or non-sensitive, depending on the type of business and the type of transaction. In any case, the sensitive and classified data must be stored and managed in a "regulated zone". With sophisticated encryption and key management, cloud storage platforms can qualify as a legitimate solution for storing and maintaining such data.

[5]https://cordis.europa.eu/project/rcn/210222/factsheet/en

selectively build the appropriate cyber-security solution that matches their needs, through combination of the correct bundles.

4.2 Related Work and Background

The increasingly connected world of people, organizations, and things is driven by the vast proliferation of digital technologies. This fact guarantees a promising future for cyber-security companies but poses a great threat for SMEs. According to Symantec [4], 60% of targeted attacks in 2015 aimed at small businesses, while "more than 430 million new unique pieces of malware were discovered". According to FireEye [5], 77% of all cybercrimes target SMEs. Simple endpoint protection through antivirus has become by far inadequate, due to the complexity and variety of cyber-threats, as well as the integration of multiple digital technologies in business processes, even in small enterprises. Modern cyber-security solutions for businesses, which are designed to provide multilayer proactive protection, use heuristics and threat-intelligence technologies to detect unknown threats, protecting a wide range of devices (e.g., PCs, servers, mobile devices, etc.) and business practices (e.g., BYOD, remote access, use of cloud-based apps and services, etc.). Due to this complexity, no single security solution can effectively address the whole threat landscape. Threats may range from relatively harmless, abusive content (such as spam messages) and other low-impact opportunistic attacks, to very harmful (malicious code), while they can escalate to targeted attacks (e.g., spyware, denial of service, etc.), with major operational and economic consequences for the enterprise.

According to ENISA [6] the top-5 threats in 2016 are mainly network-based. Consequently, a cost-effective solution for such threats could prove decisive for the future of SMEs and cannot be provided by one of the traditional methods.

Social engineering is another typical form of threat. This can be manifested either by a deceptive e-mail, installation instructions for a "free" or even "trial" piece of software, bogus sites, etc.

Moreover, Internet of Things (IoT) applications, such as healthcare and assistive technologies promise a higher level of quality of life for citizens around the world; on the other hand, however, they increase the attack surface, considerably. Legacy systems, implantable devices, and wireless networks are also eligible attack domains. Embedded systems are used more and more, e.g. in modern cars. Controlling and manipulating such entities can provide attackers with enormous power. The same holds for critical infrastructures and drones. Therefore, the cyber security research community, needs to address those issues.

SMEs consist of diverse businesses that usually operate in the service, manufacturing, engineering, agroindustry, and trade sectors. SMEs can be innovative and entrepreneurial, and usually aspire to grow. Nevertheless, some stagnate and remain family owned. There is no single, uniformly-accepted definition of SMEs. Many definitions exist whereby SMEs are classified by different characteristics, including, but not limited to profitability, turnover, sales revenue, or the number of people employed.

The European Union defines an SME combining the number of employees, along with revenue and assets. A medium-sized enterprise [7], is defined as an enterprise which employs fewer than 250 persons and whose annual turnover, does not exceed €50 million or whose annual balance-sheet total does not exceed €43 million.

SMEs represent the "middle class" of entities using computers, with single or home users at the bottom of the hierarchy and large companies or organizations at the top. As such, SMEs lack the resources typically available in the case of large organizations, while, at the same time, they need continuous and secure operation of their systems in order to function. Security can be quite expensive and since low-investment consequences on it are not evident until a significant incident takes place, it is often very tempting to allocate the minimum of resources for it.

However, a significant security incident can prove fatal for an SME, either directly (e.g., cessation of business transactions) or indirectly (e.g., bad reputation causing most of the customers to walk away or litigation). Most SMEs do not consider themselves as having data that is of interest to cyber-criminals and quite often dismiss the need for adequately addressing vulnerabilities in their infrastructure. In reality, the opposite is true; every enterprise today collects data on employees, clients, and vendors that are of interest to cyber criminals. Consequently, it is crucial to develop cyber-security products that would focus on the needs of SMEs. Challenges for mitigating cyberthreats must be addressed and highlighted, and the need to mitigate the identified risks must be addressed as well.

FORTIKA aims to establish a reliable and secure business environment for SMEs, that will provide and ensure business continuity. The FORTIKA solution is composed of modules that are designed to provide a cohesive and cost-effective set of services that address those issues. These modules are described below.

4.3 Technical Approach

This section presents the high-level deployment diagram (see Figure 4.3) of the FORTIKA modules in the two main FORTIKA systems, namely the Cloud and the SME. In the Cloud, the Marketplace, its Dashboard, and Cloud platform related modules (i.e. Orchestrator, Cloud security, Cloud Storage) are deployed; further to that, several constituent components of FORTIKA cyber security appliances (i.e. ABAC, SEARS, Encrypted data search engine, redBorder Manager) are also deployed there. At the SME level, there are two distinct cases of deployment. In the first case, the deployment is performed at the FORTIKA-GW where the GW's operational modules (depicted in red colour) and the FORTIKA security modules (lightweight modules in the

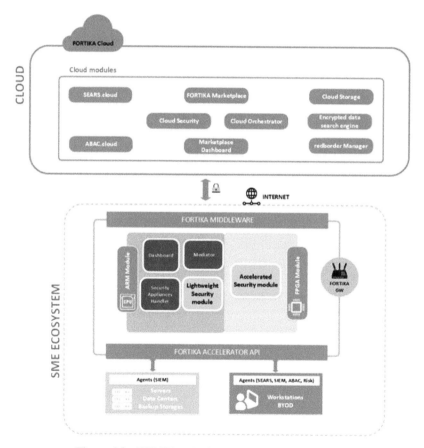

Figure 4.3 FORTIKA deployment diagram (High level).

ARM, heavyweight modules in the FPGA) can be found. In the second case, the Agents (software units collecting information and forwarding it to the GW's cyber-security appliances for processing/analysis) are deployed in the workstations and servers of the SME.

4.3.1 FORTIKA Accelerator

FORTIKA Accelerator: The FORTIKA security accelerator (FORTIKA gateway) is connected and offers unlimited expandability (by simply connecting as many accelerators in series) in terms of processing power and storage capacity, and scalability through a modular connection of two or more accelerators. Its user interface guides the enterprise administrator to appropriately define and configure the company's security & privacy policy, along with the level of encryption (information classification) and the corresponding data availability (privacy) within the enterprise and 3rd parties (e.g. suppliers, partners/ collaborators, customers, other parties), thus covering a wide range of use case scenarios. The system users/admins are kept informed at any time via comprehensive visual analytics while being able to interfere with the functionality of the presented solution in an effortless and user-friendly way.

FORTIKA Accelerator Architecture: Acceleration has been a hot topic in computing for the past few years, with Moore's law and the associated performance bumps slowly crawling to a halt. Currently, most industrial leaders accept that one form of acceleration will be used to provide the compute capacity required to cope with the large flows of data being created in the modern, widely interconnected world. FORTIKA, leverages acceleration in the form of programmable logic devices (FPGA-enabled gateway), to deliver high-performance security applications to SMEs. FPGAs offer an efficient solution in terms of performance, flexibility and power consumption. To achieve this, the FPGA must be made accessible as a resource over the network while allowing users to remotely deploy resource-demanding compute tasks on the device. This requires a middleware, either in software or in hardware to allow for the discovery of the programmable logic resources and the exchange of information between the marketplace, which is responsible for determining the appropriate infrastructure for the deployment of a task, and the accelerator module in order to determine where tasks should be deployed.

The FORTIKA accelerator module (Figure 4.4) utilizes an FPGA SoC embedded device which combines ARM processors with programmable logic in one integrated circuit. This device allows an optimal division of labour

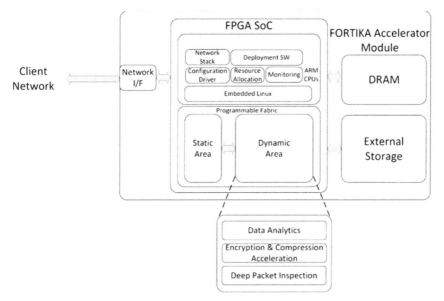

Figure 4.4 FORTIKA accelerator architecture.

between software and hardware and allows system designers, to offload computationally intensive tasks to the hardware while using the software for any light-weight, non-critical issue. FORTIKA has inherited several features from the T-NOVA FPGA-powered cloud platform, which uses OpenStack running on the CPUs to deploy tasks on the programmable logic but extended and adapted the platform to meet FORTIKA's edge demands.

The FORTIKA Middleware (MDW) (Figure 4.5) aims to facilitate a) the interactions between the FORTIKA GW and the FORTIKA marketplace; b) the loading of the security bundles to the FORTIKA accelerator; c) the exchange of data between the ARM deployed security bundles and their FPGA deployed counterparts; and d) the SW developers in producing accelerated security bundles that can be deployed in the FORTIKA accelerator. To put things in context, the following picture shows which (sub)systems, the MDW (pink Note boxes) aims to "glue" and what activities to facilitate, inside the FORTIKA architecture.

To achieve these objectives the MDW consists of several components namely the Security Bundle Handler (SBH), the LwM2M client, and the Synthesis engine. The first one provides the deployment and management of the bundles in the FORTIKA GW (both in the ARM and the FPGA parts). The second one provides the communication engine/channel which is used

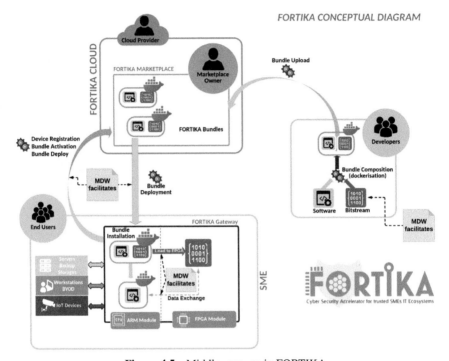

Figure 4.5   Middleware use in FORTIKA.

to interact with the FORTIKA marketplace, whereas the last alleviates the development of accelerated security bundles by hiding the complexity of HW design and configuration from the FORTIKA SW developers. As Figure 4.6 indicates, the first two components are deployed in the FORTIKA Accelerator (GW), whereas the last one is currently deployed in a Virtual Machine located at FINT's cloud infrastructure. So far, the Synthesis engine and the GW's MDW components (SBH and LwM2M client) do not have any interaction as their activities are under different scopes.

Developing applications for the FPGA requires knowledge of the HW platform and its specifics, something that can discourage SW developers from building applications for the FORTIKA accelerator. In the project's context, we tackle this issue by exploiting the fact that the FPGA application development is divided in two phases, namely the Front-End design and the Back-End design [8]. For the Front-End design phase, the FORTIKA developers are using High Level Synthesis tools (i.e. Vivado HLS suite [3]) (Figure 4.7) which allows them to write their FPGA applications in high level languages,

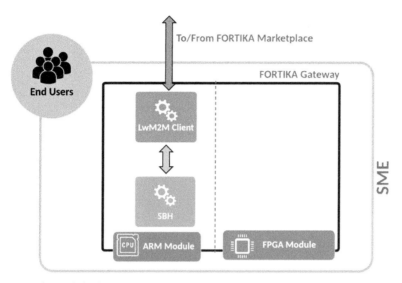

Figure 4.6 SBH and LwM2M client components of the middleware.

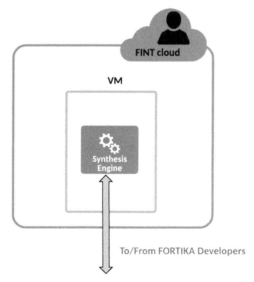

Figure 4.7 Synthesis engine component of the middleware.

such as C/C++, thus avoiding to use low level hardware specific languages (e.g. VHDL) that require knowledge of the HW specifics. After writing their code, the developers can use Vivado HLS (Figure 4.8) to produce artefacts

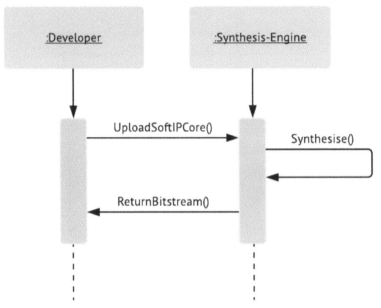

Figure 4.8   Synthesis sequence steps.

that are known as Soft IP (Intellectual Property) cores. These IP cores are used in the Back-End design phase for producing the final bitstreams, that can run on the actual FPGA; however, the Back-End design phase requires the knowledge of specific parameters of the used HW design, thus making it a hard task for the standard SW engineers; therefore, it is this design phase that the FORTIKA MDW aims to facilitate by providing a service that takes as input the produced soft IP core, runs the low-level synthesis (process of the Back-End design phase), and then returns to the developers the final bitstream. In this context, the following diagram depicts, the sequence of steps that are followed from the Synthesis Engine for implementing this task.

The *UploadSoftIPcore()* represents the function that allows developers to upload the produced soft IP cores to the Synthesis Engine. Currently, these IP cores are received via email, however at the next versions of the MDW the cores will be uploaded via a web form; this web form is planned to be provided from the Marketplace dashboard. The *Synthesise()* function, performs the low-level synthesis that produces the final bitstream. The *ReturnBitStream()* function, represents the push of the synthesised bitstream to the developer.

4.3.2 Fortika Marketplace

To facilitate competition and support different value chain configurations, a novel Marketplace Platform is introduced, allowing FORTIKA users to interact with Service Providers and multiple third-party Security Function Developers, for selecting the best service bundle that suits their needs. For this reason, the Marketplace incorporates a prototype that aims to introduce and promote a novel market field for security services, introducing new business-cases and considerably expanding market opportunities by attracting new entrants to the cyber-security market. SMEs and academia can leverage the FORTIKA architecture by developing innovative cutting-edge Security Functions, that can be included in the Function Store, and rapidly introduced to the market, thus avoiding the delay and risk of hardware integration and prototyping. By utilizing a common web-based graphical user interface, the Marketplace constitutes the environment where customers can:

- Place their requests for FORTIKA services and declare their requirements for the corresponding security functions
- Receive offerings and make the appropriate selections, considering the offered Service Level Agreements (SLAs)
- Monitor the status of the established security services and associated security functions, as well as perform, according to their rights, management operations on them (Service monitoring and management will be enabled via a graphical Service Dashboard to be implemented)

The overall concept for security functions trading, deployment and management within the Marketplace is depicted in Figure 4.9, where third-party Security Function developers (1) advertise their available virtual security appliances and users may acquire them for customized service creation/utilization. More specifically, users' requests (2) are received via the Brokerage Module as part of the Marketplace Platform, which is responsible for a) analysing their requirements, b) matching the analysis results with the available resources, maintained by the "Management & Orchestration" module along with the Security Functions aggregated at the Store (4), and c) initiating an auction process for all valid solutions under various merchandise policies and the available SLA models. Upon successful SLA establishment and Functions trading, the Orchestration module deploys the Security Function onto the underlying infrastructure (5), maintaining its control, customization and administration.

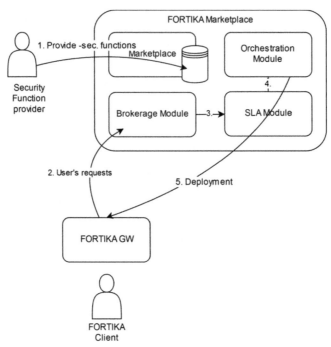

Figure 4.9   Process of deployment and management within FORTIKA marketplace.

To carry out Security Functions discovery provided by third-party developers, similarity-based algorithms such as Nearest Neighbour will be exploited by the Brokerage module to perform service matching. To speed up this process, FORTIKA will study and identify the most appropriate data structures for establishing a competent resource and service description schema for Security Functions matching the brokerage. A principal target is to identify mandatory and optional fields within the schema so as to allow a configurable degree of exposure of resources and services, associated Security Functions and SLAs to all involved actors, according to the confidentiality requirements of each. The integration of the FORTIKA Middleware appliance in existing networks requires seamless connectivity, according to usability and automation standards and guidelines. The appliance will integrate an OpenFlow Ethernet switch with physical Ethernet ports routing and security capabilities (firewall, IPS, IPSec). The appliance will also provide the required processing and storage to enable applications available through

FORTIKA Marketplace to be locally deployed but orchestrated according to rules computed in the cloud. The FORTIKA Marketplace will enable service providers to deploy and promote integrated security services through a web-based user-friendly interface with personalization features. Depending on the service design requirements, the FORTIKA Marketplace will be deployed in the cloud. Deployment of the Marketplace is not limited to public or private cloud. Due to the dynamical deployment mechanisms leveraging tools like Ansible and Docker, and the use of standards (TOSCA) for the services definition, FORTIKA consortium is not limited to any type of cloud resources.

FORTIKA Appliances (Virtual or Physical) will be managed through a FORTIKA-specific management network, using a personalized cloud service. For this reason, an integrated management platform will be deployed which will offer a consistent and unique administrator front end, for both the Middleware appliance configuration as well as installed modules configuration and management. The administrator front end, will allow management of the Security Functions' lifecycle.

Finally, the connection of FORTIKA Middleware appliances with the orchestrator in the cloud, is a critical point since protecting the integrity and confidentiality of data traveling in the fog area is crucial for middleware adoption and end-user trust to FORTIKA. For this reason, FORTIKA Middleware and FORTIKA cloud services communicate over secure channel leveraging LWM2M protocol. This is the back-channel used for management of the FORTIKA Appliance with the running Middleware.

4.4 Indicative FORTIKA Bundles

4.4.1 Attribute-based Access Control (ABAC)

Access control can be defined as a security service, co-existing with others, that aims to limit actions or operations of legitimate entities against requested resources [9]. Over the years, many access-control models have been proposed with the prevalent ones being MAC, DAC and RBAC [9]. In the recent years, information systems are able to interact with the environment, the context, thus a need for a novel approach in controlling access on context-aware information systems arose. As a result, Attribute-Based Access Control (ABAC) was proposed. ABAC policies are able to include attributes of the subject (requestor), the object (requested resource) and the context (environment). So, in contrary to legacy models, based on identities, a higher level of versatility and control can be achieved.

FORTIKA implements ABAC by providing a cloud-based access control solution which will be highly benefited from the FORTIKA Gateway appliance, to control access to SME ecosystem resources, based on policies that the SME will be able to create and manage.

A system that implements ABAC, consists of the following components [10] (Figure 4.10):

- Policy Administration Point (PAP), that is used to create, store, test and retrieve access control policies. Since the PAP component will be hosted in FORTIKA cloud, a multi-tenant environment will be deployed so that SME administrator users will have access to own organization policies only.
- Policy Information Point (PIP), that retrieves all necessary attributes and authorization data required by PDP in order to reach an access control decision. PIP in FORTIKA is implemented twofold both in the cloud and in the fog, since attribute values are collected from both the cloud and from SME premises.
- Policy Decision Point (PDP) that evaluates access requests against policies so that access control decision is computed.

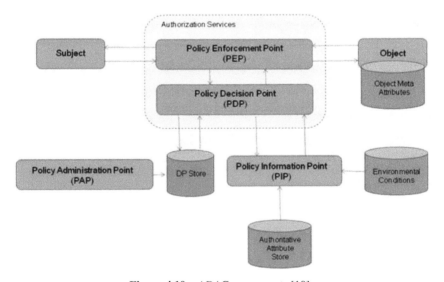

Figure 4.10   ABAC components [10].

- Policy Enforcement Point (PEP) which is the component where an access control request is generated and access decision is enforced.

The Fortika ABAC service is designed as a three-layered approach (Figure 4.11). In terms of component placing and communication architecture, the PIP and PAP components, as well as the related Policy Repository, will be deployed in the cloud (ABAC.Cloud). This will allow for rapid policy replication in case of multi-site SMEs and, additionally, will permit for replacing an on premise FORTIKA appliance without any prior consideration for existing attributes and policies. Moreover, cloud can provide adequate processing and storage resources to create a user-friendly administration environment.

On the other hand, to avoid any issues with network latency or network unavailability [11], the PDP component will be held in the fog area (ABAC.fog). More specifically, PDP will be held in FORTIKA's physical or virtual appliance hosted in SME premises, thus accelerating decision making. Additionally, to better support contextual attributes, a local PIP along with a local attribute repository (currently labelled NA-PIP) will accompany PDP and communicate with cloud PIP to exchange attribute information.

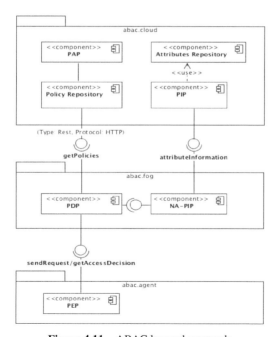

Figure 4.11   ABAC layered approach.

Finally, the PEP component will be initially integrated into a prototype agent for client devices. Nevertheless, ABAC solution will provide the appropriate API, for other compatible PEP components to be able to utilize FORTIKA's ABAC service.

FORTIKA ABAC implements the XACML framework [12] and is based exclusively on open-source technologies, developed with Java and Java EE using Maven. ABAC.Cloud is based on WSO2 Identity Server which is licensed under Apache 2.0 license, whereas ABAC.fog is based on Balana XACML and has been developed to provide a RESTful API to PEPs. The API exposes services according to OASIS REST Profile for XACML 3.0 version 1.0 [13]. This enables potentially any vendor or integrator to utilize FORTIKA ABAC.fog and consume authorization services, constituting FORTIKA ABAC an Authorization as a Service (AuthZaaS) offering.

4.5 Social Engineering Attack Recognition Service (SEARS)

Social engineering attacks are usually an important step in the planning and execution of many other types of cyber-attacks. The term 'social engineering' refers to physiological, emotional and intellectual manipulation of people into performing actions or revealing confidential information. As defined in [14], social engineering is: "*a deceptive process whereby crackers 'engineer' or design a social situation to trick others into allowing them access an otherwise closed network, or into believing a reality that does not exist.*"

The increased usage of electronic communication tools (email, instant messaging, etc.) in enterprise environments results in the creation of new attack vectors for social engineers. However, a successful social engineering attack could result in a compromised SME's information system. Thus, several attempts have been made in the research field to provide technical means for detecting such attacks in early stages. Works that are near to a prototyping level are SEDA [15] and SEADM [16]. Furthermore, interesting efforts that are still under development in the research laboratory are [17] and [18].

Social Engineering Attack Recognition System (SEARS) will operate in the application layer and will be able to compute communication risk and therefore prevent personal or corporate data leakage by raising alerts to the employees when the chat conversation reaches a specific risk threshold [19]. SEARS is a collection of autonomous services that collaborate with each other through technology-agnostic messaging protocols, either point-to-point or asynchronously. The development of SEARS components follows the microservices design approach. Namely, each component is consisted of a number of independent microservices that serve distinct functionalities of the whole system.

SEARS components will be placed in the three layers of FORTIKA's architecture, as follows:

Client layer:

The SEARS Agent (SEARS.agent) is a service that monitors, captures and pre-processes an employee's social media communications. It is also capable of receiving the total risk value and alerting the user for possible social engineering attack attempts. SEARS.Agent is deployed on end-user's device in a form of a docker container or as local service and continuously monitors and captures an employee's social media communications. SEARS users are registered SME employees as interlocutors (e.g. working on live chat service) or corporate IT administrators (Figure 4.12).

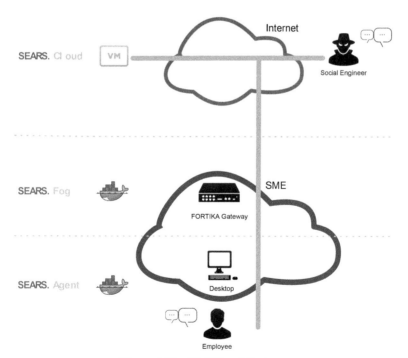

Figure 4.12 SEARS architecture.

Fog layer:

SEARS components in the fog area (SEARS.fog) will be deployed in FOR-TIKA physical or virtual appliance, hosted in SME premises. SERS.fog receives the captured data and stores it (Detection Storage component) locally for further pre-processing (Pre-processing component), using Natural Language Processing techniques. The pre-processed data is then anonymized and sent to the cloud (SEARS.cloud). The Detection Engine receives the particular risk values from the SEARS.cloud and then calculates the total Social Engineering Risk value, stores it in the Detection Repository and sends it to the SEARS.client. SEARS.Fog is deployed on FORTIKA Gateway in the form of a docker container.

Cloud layer:

The pre-processed data received from the SEARS.fog is stored in the SEARS Storage component of SEARS.cloud, in order to be used by the Risk Estimation component to calculate values of particular risks. These values are then sent to the SEARS.fog. The following components are part of SEARS.Cloud core functionality and implemented by several microservices.

- **Document Classification (DC):**
  Text dialogue, in the form of an anonymized TF-IDF matrix, is processed and classified as dangerous or not. The real text dialogue is processed at the SEARS.Agent, where an anonymized frequency vector is delivered to SEARS.Cloud, where the classification takes place.

- **Personality Recognition (PR):**
  Each of the interlocutors is being classified based on his/her writings. The processing/classification takes place at the SEARS.Cloud using the previous anonymized frequency vector.

- **User History (UH):**
  Each previous text chat between the two specific interlocutors is represented as probability (decimal number) and it is stored at SEARS.Cloud.

- **Exposure Time (ET):**
  The duration of an employee's online presence is being depicted as a decimal number stored at SEARS.Cloud

SEARS offers the ability to communicate the estimated risk values to other modules of FORTIKA. The outgoing information is provided using a standard HTTP POST method. All data is encoded using the JavaScript Object Notation (JSON) format and follow the structure of SEARS Output JSON Schema. Moreover, all data transfers are being carried out using REST

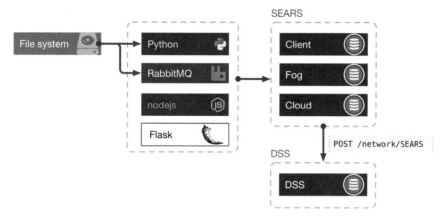

Figure 4.13 SEARS conceptual design.

APIs through HTTPS protocol thus the communication channel cannot be compromised. The SEARS conceptual design as a whole is presented in Figure 4.13.

SIEM

The Security Information and Event Management System (SIEM), is a solution able to analyse information and events collected at different levels of the monitored system in order to discover possible ongoing attacks, or anomalous situations. FORTIKA includes a customized SIEM solution, able to deal with specificities of its different technologies and components.

The network provides real-time traffic data to the SIEM system. The system in turn, forwards the data for processing to both the Anomaly detection and the behavioural analysis components. The Anomaly detection component analyses the data in order to detect anomalies, utilising both automatic anomaly detection algorithms, such as Local Outlier Factor and Bayesian Robust Principal Component Analysis, as well as visual analytics methods, such as k-partite graphs and multi-objective visualizations. The Behavioural analysis component processes the network data in order to identify abnormal traffic patterns that may indicate that a malicious event such as a DDoS attack is in progress. The output from both components is then passed to the Visualization component for presentation to the user, or to the Hypothesis Formulation component. The Hypothesis Formulation component performs a statistical analysis of the output data of the Anomaly detection and the

Behavioural analysis component, through a series of hypotheses in order to determine whether these data express a normal or a usual traffic pattern or behaviour. The analysis data can be subsequently fed back to the Anomaly detection and the behavioural analysis components for further analysis.

4.6 Conclusion

FORTIKA architecture proposes a hybrid (hardware software) cybersecurity solution suitable for micro, small and medium-sized enterprises allowing them to continuously integrate novel cyber-security technologies and thus reinforce their position and overall reputation in the European market. Concluding, this paper introduced a novel architecture that aims at reshaping the cyber-security landscape in order to provide an end-user-friendly solution targeting towards moving security near the network edge. This architecture is based upon two pillars: A near-the-edge security-accelerator, which is able to "accelerate" security in the place where the problem is formulated, and a Cloud Marketplace which provides a unified portal for enabling security for FORTIKA end-users. The preliminary evaluation of the presented work illustrated that users (SMEs) can identify which cyber-security solutions are suitable for their enterprises and seamlessly deploy them on their infrastructures (FORTIKA gateway). Additionally, security-solutions' developers/providers can easily offer their services through the FORTIKA marketplace, which also allows them to interact with users and offer custom-tailored cyber-security solutions (brokerage), thus extending their marketing opportunities. The presented work is an ongoing EU-funded Horizon 2020 project, and currently runs the second year of development. Several complex and intuitive features are to be developed in the near future and thus more detailed and elaborate reporting of the work will be presented through publications and public workshops, as well as from the project's social media accounts (Facebook, Twitter, YouTube, etc.).

Acknowledgment

This work has received funding from the European Union's Horizon 2020 Framework Programme for Research and Innovation, with Title H2020-FORTIKA "cyber-security Accelerator for trusted SMEs IT Ecosystem" under grant agreement no. 740690.

References

[1] H. Madsen, G. Albeanu, B. Burtschy, and F. Popentiu-Vladicescu, "Reliability in the utility computing era: Towards reliable fog computing," in International Conference on Systems, Signals, and Image Processing. IEEE, jul 2013, pp. 43–46. [Online]. Available: http://ieeexplore.ieee.org/document/6623445/

[2] Y. Nikoloudakis, S. Panagiotakis, E. Markakis, E. Pallis, G. Mastorakis, C. X. Mavromoustakis, and C. Dobre, "A Fog-Based Emergency System for Smart Enhanced Living Environments," IEEE Cloud Computing, vol. 3, no. 6, pp. 54–62, nov 2016. [Online]. Available: http://ieeexplore.ieee.org/document/7802535/

[3] C. Dobre, C. X. Mavromoustakis, N. M. Garcia, G. Mastorakis, and R. I. Goleva, "Introduction to the AAL and ELE Systems," Ambient Assisted Living and Enhanced Living Environments: Principles, Technologies and Control, pp. 1–16, jan 2016. [Online]. Available: https://www.sciencedirect.com/science/article/pii/B9780128051955000 016 Books (IDEA/IGI, Springer and Elsevier).

[4] B. McKenna, "Symantec's Thompson pronounces old style IT security dead," Network Security, vol. 2005, no. 2, pp. 1–3, 2005. [Online]. Available: http://linkinghub.elsevier.com/retrieve/pii/S135348580500, 1947."

[5] Cyber Security Experts & Solution Providers |FireEye." [Online]. Available: https://www.fireeye.com/

[6] "ENISA Threat Landscape Report 2016 — ENISA." [Online]. Available: https://www.enisa.europa.eu/publications/enisa-threat-land scape-report-2016. [Accessed: 20-Nov-2017].

[7] E. O. Yeboah-Boateng, Cyber-Security Challenges with SMEs in Developing Economies: Issues of Confidentiality, Integrity & Availability (CIA). Institut for Elektroniske Systemer, Aalborg Universitet, 2013.

[8] E. K. Markakis, K. Karras, A. Sideris, G. Alexiou, and E. Pallis, "Computing, Caching, and Communication at the Edge: The Cornerstone for Building a Versatile 5G Ecosystem," IEEE Communications Magazine, vol. 55, no. 11, pp. 152–157, nov 2017. [Online]. Available: http://ieeexplore.ieee.org/document/8114566/

[9] R. S. Sandhu and P. Samarati, "Access control: principle and practice," IEEE Communications Magazine, vol. 32, no. 9, pp. 40–48, Sep. 1994.

[10] V. C. Hu et al., Guide to Attribute Based Access Control (ABAC) Definition and Considerations (Draft). 2013.

[11] S. Salonikias, I. Mavridis, and D. Gritzalis, "Access Control Issues in Utilizing Fog Computing for Transport Infrastructure," in Critical Information Infrastructures Security, 2016, pp. 15–26.

[12] "eXtensible Access Control Markup Language (XACML) Version 3.0." [Online]. Available: http://docs.oasis-open.org/xacml/3.0/xacml-3.0-core-spec-os-en.html. [Accessed: 18-Jan-2019].

[13] "REST Profile of XACML v3.0 Version 1.0." [Online]. Available: http://docs.oasis-open.org/xacml/xacml-rest/v1.0/csprd03/xacml-rest-v1.0-csprd03.html. [Accessed: 18-Jan-2019].

[14] B. H. Schell, B. Schell, and C. Martin, Webster's New World Hacker Dictionary. John Wiley & Sons, 2006.

[15] M. D. Hoeschele and M. K. Rogers, "CERIAS Tech Report 2005–19 Detecting Social Engineering," 2004.

[16] F. Mouton, L. Leenen, and H. S. Venter, "Social Engineering Attack Detection Model: SEADMv2," 2015, pp. 216–223.

[17] R. Bhakta and I. G. Harris, "Semantic analysis of dialogs to detect social engineering attacks," in Semantic Computing (ICSC), 2015 IEEE International Conference on, 2015, pp. 424–427.

[18] S. Uebelacker and S. Quiel, "The Social Engineering Personality Framework," 2014, pp. 24–30.

[19] N. Tsinganos, G. Sakellariou, P. Fouliras, and I. Mavridis, "Towards an Automated Recognition System for Chat-based Social Engineering Attacks in Enterprise Environments," in Proceedings of the 13th International Conference on Availability, Reliability and Security – ARES 2018, Hamburg, Germany, 2018, pp. 1–10.

[20] C. Liu, Y. Mao, J. E. Van der Merwe, and M. F. Fernández, "Cloud Resource Orchestration: A Data-Centric Approach," in Proceedings of the biennial Conference on Innovative Data Systems Research (CIDR), 2011, pp. 241–248. [Online]. Available: http://www2.research.att.com/maoy/pub/cidr11.pdf

[21] A. Dubey and D. Wagle, "Delivering software as a service," The McKinsey Quarterly, vol. 6, no. May, pp. 1–12, 2007. [Online]. Available: http://www.pocsolutions.net/Delivering_software_as_a_service.pdf

[22] K. Lane, "Overview Of The Backend as a Service (BaaS) Space," 2013. [Online]. Available: http://www.integrove.com/wp-content/uploads/2014/11/api-evangelist-baas-whitepaper.pdf

[23] S. A. Fahmy, K. Vipin, and S. Shreejith, "Virtualized FPGA accelerators for efficient cloud computing," in Proceedings IEEE 7th International Conference on Cloud Computing Technology and Science, CloudCom 2015. IEEE, nov 2016, pp. 430–435. [Online]. Available: http://ieeexplore.ieee.org/document/7396187/

[24] J. A. Williams, A. S. Dawood, and S. J. Visser, "FPGA-based cloud detection for real-time onboard remote sensing," in Proceedings 2002 IEEE International Conference on Field-Programmable Technology, FPT 2002. IEEE, 2002, pp. 110–116. [Online]. Available: http://ieeexplore.ieee.org/document/1188671/

[25] S. Byma, J. G. Steffan, H. Bannazadeh, A. Leon-Garcia, and P. Chow, "FPGAs in the cloud: Booting virtualized hardware accelerators with OpenStack," in Proceedings 2014 IEEE 22nd International Symposium on Field-Programmable Custom Computing Machines, FCCM 2014. IEEE, may 2014, pp. 109–116. [Online]. Available: http://ieeexplore.ieee.org/document/6861604/

[26] L. Xu, W. Shi, and T. Suh, "PFC: Privacy preserving FPGA cloud A case study of MapReduce," in IEEE International Conference on Cloud Computing, CLOUD. IEEE, jun 2014, pp. 280–287. [Online]. Available: http://ieeexplore.ieee.org/lpdocs/epic03/wrapper.htm?arnumber=6973752

[27] K. Karras, O. Kipouridis, N. Zotos, E. Markakis, and G. Bogdos, "A Cloud Acceleration Platform for Edge and Cloud," in EnESCE: Workshop on Energy-efficient Servers for Cloud and Edge Computing, 2017. [Online]. Available: https://www.researchgate.net/publication/313236609

[28] Nikoloudakis, Y, Pallis, E, Mastorakis, G, Mavromoustakis, CX, Skianis, C & Markakis, EK 2019, 'Vulnerability assessment as a service for fog-centric ICT ecosystems: A healthcare use case' Peer-to-Peer Networking and Applications. https://doi.org/10.1007/s12083-019-0716-y

5

CYBECO: Supporting Cyber-Insurance from a Behavioural Choice Perspective

Nikos Vassileiadis[1], Aitor Couce Vieira[2], David Ríos Insua[2], Vassilis Chatzigiannakis[3], Sofia Tsekeridou[3], Yolanda Gómez[4], José Vila[4], Deepak Subramanian[5], Caroline Baylon[5], Katsiaryna Labunets[6], Wolter Pieters[6], Pamela Briggs[7] and Dawn Branley-Bell[7]

[1]Trek Consulting, Greece
[2]Institute of Mathematical Sciences (ICMAT), Spanish National Research Council (CSIC), Spain
[3]Intrasoft International, Greece
[4]Devstat, Spain
[5]AXA Technology Services, France
[6]Faculty of Technology, Policy and Management, Delft University of Technology, the Netherlands
[7]Psychology, University of Northumbria at Newcastle, United Kingdom
E-mail: n.vasileiadis@trek-development.eu; aitor.couce@icmat.es; david.rios@icmat.es; Vassilis.Chatzigiannakis@intrasoft-intl.com; Sofia.Tsekeridou@intrasoft-intl.com; ygomez@devstat.com; jvila@devstat.com; deepak.subramanian@axa.com; caroline.baylon@axa.com; K.Labunets@tudelft.nl; W.Pieters@tudelft.nl; p.briggs@northumbria.ac.uk; dawn.branley-bell@northumbria.ac.uk

Cyber-insurance can fulfil a key role in improving cybersecurity within companies by providing incentives for them to improve their security, requiring certain minimum protection standards. Unfortunately, so far, cyber-insurance has not been widely adopted. CYBECO focuses on two aspects to fill this gap: (1) including cyber threat behaviour through adversarial risk analysis to

103

support insurance companies in estimating risks and setting premiums and (2) using behavioural experiments to improve IT owners' cybersecurity decisions. We thus facilitate risk-based cybersecurity investments supporting insurers in their cyber offerings through a risk management modelling framework and tool.

5.1 Introduction

Cyber security is increasingly perceived as a major global problem as reflected by the World Economic Forum [1] and is becoming even more important as companies, administrations and individuals get more and more interconnected, facilitating the spread of cyberthreats. Famous examples include the Target 2014 data breach, in which a cyber attack to that company through one of its suppliers caused the loss of 70 million credit card details, entailing major reputational damage, and the NotPetya malware, which affected thousands of organisations worldwide with an estimated cost of more than 8 billion EUR.

Given the importance of this problem, numerous frameworks have been developed to support cybersecurity risk management, including ISO 27005 [2] or CORAS [3], among several others. Similarly, several compliance and control assessment frameworks, like ISO 27001 [4] or Common Criteria [5], provide guidance on the implementation of cybersecurity best practices. Their extensive catalogues of assets, controls and threats and their detailed guidelines for the implementation of countermeasures to protect digital assets facilitate cyber security engineering. However, a detailed study of the main approaches to cybersecurity risk management reveals that they often rely on risk matrices for risk analysis purposes, with shortcomings documented in e.g. Thomas et al. [6].

Moreover, with few exceptions like IS1 [7], such methodologies do not explicitly take into account the intentionality of certain threats, in contrast with the relevance that organisations like the Information Security Forum (ISF) [8] start to give to such threats. As a consequence, ICT owners may obtain unsatisfactory results in relation with the prioritisation of cyber risks and the measures they should implement, even more in the case of an increasing variety of threats as well as the increasing complexity of countermeasures for risk management available, including the recent emergence of cyber-insurance products [9].

The CYBECO project aims at providing a framework and a tool to facilitate cyber security resource allocation processes, including the provision of cyber insurance and, consequently, contribute to a more cyber secure environment.

5.2 An Ecosystem for Cybersecurity and Cyber-Insurance

CYBECO includes a detailed analysis of the cyber-insurance (and cyber-security) ecosystem. This is aimed at facilitating the use of the toolbox for specific stakeholder scenarios, as well as providing policy recommendations that, together with the toolbox, help achieve key goals. We identified several primary and secondary actors participating in the cyber-insurance ecosystem and relationships that exist between them.

The main parties that we identified are:

- *insurance providers* who "assume risks of another parties in exchange for payment" [9];
- *insurance brokers* who provide an advice to the companies on the available insurance products matching their needs;
- *companies* that are interested in transferring part of their cyber-related risks with cyber-insurance. The reasons for purchasing cyber-insurance may differ depending on the company size.

Secondary actors include *consumers* using services or products provided by companies; *experts* that provide professional services to the insurance companies (e.g., risk assessment, forensics, cyber incident counsel, legal and PR services); *regulators* managing corresponding business sectors; and other parties.

Based on the discussions with the representatives of different actor types and existing literature, we identified their motivation and goals, which guide their behaviour in the ecosystem. An insurance provider is interested in increasing its market share, having better actuarial data to improve risk assessment and run a profitable business. Similarly, an insurance broker aims at making a profit, but also at providing its clients with high-quality advice about cyber risks. The companies try to get advice on security investments, cover possible losses related to cyber risks and, in case of an incident, get help with incident handling. At a higher level, we have a regulator or government actor whose primary interests are to increase the overall level of security and create a resilient ecosystem [10].

The current cybersecurity regulations and standards are poor concerning policy measures that are related to cyber-insurance. Therefore, we adopted a framework proposed by Woods and Simpson [10] to identify possible policy measures that can be considered by the government for improving the cyber-insurance market. The framework provides six main themes for possible policy measures:

1. *Wider adoption* covers measures like assigning financial costs to cyber events (i.e., regulatory fines), raising awareness that traditional insurance policies do not cover cyber risk, supporting market development via governmental procurement capability, and making cyber-insurance mandatory for specific business sectors.
2. *Defining coverage* includes standardisation of the language used in cyber-insurance policies, promotion of cyber exclusion clauses in non-cyber policies, and providing certification for acts of cyber war or terrorism.
3. *Data collection* includes policy measures such as the introduction of standard data formats for risk assessment and claim processes, requirements for risk assessment data collection, and collecting high-level data on the cyber-insurance market.
4. *Information sharing* consists of measures like making available data collected by government (related to GDPR or NIS regulations), open access to sector-specific information-sharing initiatives (sector ISACs), creating a state- or EU-level cyber incident data repository and mandating other organisations to share data.
5. *Best practice* includes defining cybersecurity best practices that cyber-insurers should check with their clients or even demand and, at the same time, implementing regulations that clarify what the liability of insurers giving security advice is.
6. *Catastrophic loss* comprises policy measures related to the role of government as insurer of last resort, including different models for insuring catastrophic events (e.g. terrorism).

To better understand which policy measures have more influence on the ecosystem, we mapped the goals of the actors to Wood and Simpsons' framework. Wider adoption of cyber-insurance implies growth of the market and, therefore, supports goals like increasing market share for insurers, making a profit for insurers and brokers. At the same time, wider adoption means that more companies insured their cyber risks, implying that the resilience of the ecosystem is also increasing. Policy measures related to coverage definition help brokers to better advice companies about relevant insurance products meaning that companies get an appropriate policy to cover their cyber risks. Wider use of cyber exclusions in non-cyber policies could lead to improving the level of sales of cyber-insurance products contributing to the profitability of insurers and brokers.

Data collection policy measures impact insurers' goal related to having better actuarial data. Information sharing measures also supply insurers with actuarial data and help brokers to provide clients with high-quality advice about cyber risks as brokers can have real information about current cyber incidents. Security best practices help brokers to advise their clients on cyber risks and countermeasures, meaning that companies get advice about what security investments to make. By using security standards in cyber-insurance risk assessment and even including security best practices as required in cyber-insurance policy, the government could affect the overall level of security in the ecosystem. Finally, catastrophic loss measures contribute to increasing ecosystem resilience, which is the goal of the governmental actor.

The only goal that is not covered by this policy measures framework is related to company actors who need assistance in incident handling. However, the existing practice shows that most insurers offer their clients crisis management services as a part of cyber-insurance products. Such services are mostly provided by partnering organisation and its cost is included in the policy coverage [11, 12].

Details on the cyber-insurance ecosystem, the associated policy recommendations, and their connection with the CYBECO toolbox are described in the associated deliverable [14].

5.3 The Basic Cybeco Model: Choosing the Optimal Cybersecurity and Cyber-Insurance Portfolio

CYBECO provides several cyber-insurance related decisions. The main model aims at providing support to an organisation that needs to allocate its cybersecurity resources, including the adoption of cyber-insurance. In it, we distinguish between a Defender, to which our methodology will support in her allocation, and an Attacker, who will try to perpetrate attacks to the Defender in pursue of certain goals.

We represent the problem as a bi-agent influence diagram (BAID) in Figure 5.1, with the terminology used in [14]. Therefore, the diagram includes oval nodes that represent uncertainties modelled with probability distributions; hexagonal utility nodes that represent preferences modelled with a utility function; rectangle nodes, which represent decisions modelled through

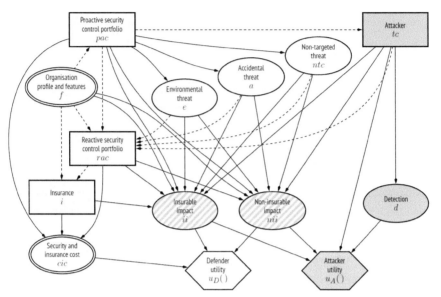

Figure 5.1   BAID describing the cybersecurity resource allocation problem.

the set of relevant alternatives at such point; and, finally, double oval nodes that represent deterministic nodes modelled through a function evaluating the antecessors of the corresponding node. The diagram also includes arrows to be interpreted as in standard influence diagrams [15]. Light nodes designate nodes belonging just to the Defender problem; dark ones to the Attacker; and, finally, striped ones are relevant to both agents.

We outline the BAID. First, we include a description of the organization profile and features, including its assets. We then identify the threats relevant to the organisation; following the ISF classification, we distinguish between environmental, accidental and non-targeted cyber threats, which we model through uncertain nodes. Besides, we also consider targeted cyber threats, modelled as decisions, but associated with a different agent, the Attacker. Having determined the threats and relevant assets, we may identify the impacts that we separate between insurable and non-insurable ones.

Once with the relevant threats and impacts for the organisation at hand, we may identify the actions that may be undertaken to mitigate the likelihood and/or impact of the threats. We distinguish three types of instruments: proactive security controls, reactive security controls and insurance. The above instruments may have to satisfy certain constraints (financial, technical, compliance, etc.). Besides, they will have security and insurance costs, which will typically be deterministic. With all the relevant attributes in place, we may then prepare the preference model for the Defender through her utility.

We turn now to the remaining elements of the Attacker problem, mainly his detection and identification. Finally, with all his relevant elements in place, we may then build a preference model for the Attacker through the utility of the attacker through a value node.

Based on such model, we build the so-called Defender problem. This facilitates the quantitative modelling of the problem using conditional probability distributions at uncertain nodes and a utility function for modelling the preferences and risk attitudes of the Defender. All those models are standard in decision analysis except those referring to the likely threats performed by the attacker(s) that entail strategic thinking.

To facilitate their assessment, we consider the so-called Attacker problem. As we do not have full access to the attackers to elicit their beliefs and preferences, we use random probabilities and utilities to model our uncertainty about them. We then simulate from such problem to find the corresponding random optimal alternatives that help us to find the required attack forecasts. This feeds back the Defender problem that is finally solved to provide the optimal proactive portfolio, reactive portfolio and insurance that should be implemented by supported organisation.

This and other models for other cyber-insurance related decisions are fully described in [17].

5.4 Validating CYBECO

The findings of the CYBECO project have been validated in several ways:

1. A set of use cases and scenarios were developed to verify whether the proposed models were robust in all situations. They are available in [18]. They have confirmed the validity of our approach, although some fine tuning, specification and further modelling has been required.
2. A workshop in which we presented the CYBECO toolbox wireframes to a number of cybersecurity professionals and solicited their feedback. This was essential for the fine-tuning of the project findings.
3. The last validation approach focused on the application of behavioural-experimental methods to test the assumptions of the CYBECO models on purchase behaviour of cyber-protection measures and cyber-insurance, as well as on the belief formation of cyber-risk and vulnerability levels. To this end, the project has designed and run a large-scale

online behavioural economic experiment with a total sample of 4.800 subjects from Germany, Poland, Spain and UK. Beyond the validation of the model, the experiment has provided behavioural insights relevant for the development of the cyber-insurance market in the EU.

The structure of the experiment was as follows. In a controlled gamified environment, subjects were meant to design the protection and cyber-insurance strategy for an SME and were required to carry out certain tasks online (see Figure 5.2). After that, each subject may receive a random attack

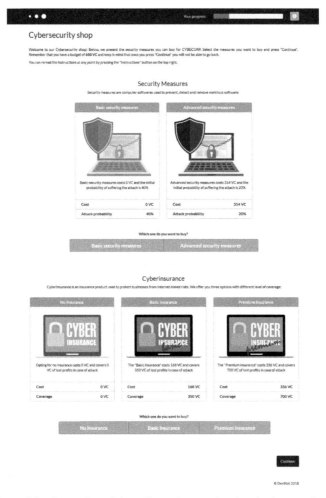

Figure 5.2 Screenshot of the online cybersecurity shop in the experiment.

with success probability depending on the purchased protection measures and level of security of her online behaviour. According to the methodology of behavioural economics, the decisions of the participants and the random events in the experiment (the attack) have an actual impact in their economic incentives, to be received after completing the experiment. To check belief formation, the process is repeated twice. The experiment also included a questionnaire to measure risk attitude and the Protection-Motivation psychological variables.

The economic experiment validated the underlying assumptions of the model and provided other relevant insights. Experimental results showed that belief formation is dependent on the context of the attack, the participants selecting higher protection and insurance levels under the menace of intentional attacks (cybercrime) than of random (random virus) ones. The experiment also analysed the impact of the experience of suffering a cyberattack in the updating of beliefs and protection-insurance strategies. The results show the presence of two opposite reactions: although an attack does in general motivate participants to increase their protection levels, suffering the attack reduced confidence level in the effectivity of the protection measure for 15.1% of the participants who reduced their protection level after the attack. As insurance behaviour regards, experimental subjects seem to purchase insurance levels over the optimal level. Moreover, the experiment excluded moral hazard in cyber-insurance: purchasing a cyber-insurance policy does not reduce the security level of online behaviour and is positively correlated with the acquisition of stronger cybersecurity protection measures. An additional relevant result of the experiment is the existence of vulnerable segments of population (elder citizens, for instance) that, although being risk averse and concerned with cybersecurity, behave insecurely online. The likely reason for this lack of security is that they do not know how to behave in a safer way.

5.5 The CYBECO Decision Support Tool

When compared with standard approaches in cybersecurity, the CYBECO paradigm provides a more comprehensive method leading to a more detailed modelling of cyber risk problems, yet, no doubt, more demanding in terms of analysis. We believe though that in many organizations, especially, in critical infrastructure sectors, the stakes at play are so high that this additional work should be worth the effort.

To facilitate implementation, we are converting our generic actionable model into a decision support system (DSS), the CYBECO tool, for cybersecurity risk management at a strategic level. The objective of such DSS would be to provide the best portfolio of security controls and insurance products, given a predefined relevant budget and other technical and legal constraints for a certain planning period.

The toolbox adopts the form of an online calculator (see Figure 5.3) to guide the user into analysing their current cybersecurity risk level and the optimal cybersecurity strategy for their specific needs. The calculator is viewed as a multi-step online visually-enriched form, which asks the pertinent

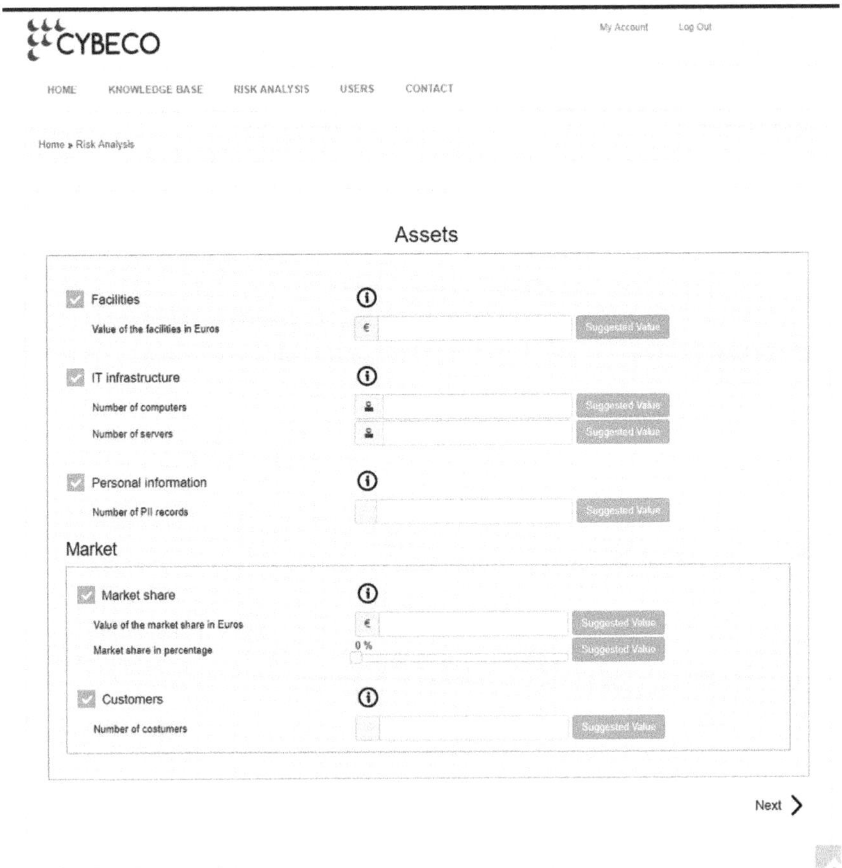

Figure 5.3 A snapshot of the CYBECO tool, gathering inputs on assets to feed the cyber risk analysis tool.

questions (e.g., company size, characteristics, relevant threats, relevant security measures and insurance products, relevant impacts, etc.) and offers the best option for the stakeholder (SME, large industry) based on the outcomes of the CYBECO cyber risk management models.

To enhance the usability, visual appearance of outputs, and general user-friendliness of the calculator, three types of user-oriented validations have been undertaken to collect relevant feedback. First, we have designed and implemented a behavioural economic experiment with a sample of 2,000 potential users of the calculator (workers in SMEs in managerial or cyber-security related positions) in Germany, Poland, Spain and UK. In a gamified controlled environment, the participants were asked to define the cyber-protection and cyber-insurance strategies of an SME using five different framings of the output of the CYBECO calculator. The experiment showed that the potential users of the CYBECO toolbox tend to use it more as an information source to make such a decision in a better informed manner rather than an expert tool able to guide them to the best option and provide relevant recommendations (only 30% of the users declared to have purchased the strategy recommended by the tool). It must be highlighted that this result is not attributable to a lack of understanding of the ranking criteria but it results from the fact that users do consciously prefer a different protection approach, coverage or price level than the one dynamically recommended by the toolbox. Another evaluation target has been the user navigation paths, offered by the toolbox, which were evaluated by two focus groups with about 50 actual users, which helped in improving the visual aspect of the toolbox. Finally, a rich set of uses cases has been developed and applied as usage patterns on the toolbox to crosscheck the correct implementation of the cyber risk analysis algorithms.

5.6 Conclusion

We have provided a brief summary of some of the ongoing and expected achievements of the CYBECO project. On the supply side, we expect that the end-users would benefit from better founded and designed cyber-insurance products and cyber risk management frameworks. On the demand side, we expect that the end-users would benefit from a well-founded tool that allows them to determine their optimal cyber security investments, including the appropriate cyber-insurance product. Globally, the society as a whole would benefit as CYBECO helps in creating a more secure environment.

In a nutshell, by properly modelling and combining decision-making behaviour surrounding cyber threats (risk generation), the decision-making behaviour of insurance companies (risk assessment) and the decision-making behaviour of IT owners (which includes cyber-insurance), we hope to help mitigate cyber risks at the global level.

Acknowledgements

CYBECO: Supporting cyberinsurance from a behavioural choice perspective is a project funded by the H2020 programme through grant agreement no. 740920.

References

[1] World Economic Forum, "The Global Risks Report 2019," 2019.

[2] International Organization for Standardization, ISO/IEC 27005 – Information Security Risk Management, 2013.

[3] M. S. Lund, B. Solhaug and K. Stølen, Model-driven Risk Analysis: The CORAS Approach, Springer, 2010.

[4] International Organization for Standardization, ISO/IEC 27001 – Information Security Management Systems – Requirements, 2013.

[5] The Common Criteria Recognition Agreement Members., Common Criteria for Information Technology Security Evaluation, Version 3.1 Release 4, 2009.

[6] P. Thomas, R. B. Bratvold and J. E. Bickel, "The risk of using risk matrices," in *SPE Annual Technical Conference and Exhibition 2013.*, 2013.

[7] National Technical Authority for Information Assurance (UK), HMG IA Standard Number 1., 2012.

[8] Information Security Forum, Information Risk Assessment Methodology 2, 2016.

[9] A. Marota, F. Martinelli, S. Nanni, A. Orlando and A. Yautsiukhin, "Cyber-insurance survey," *Computer Science Review,* 2017.

[10] PricewaterhouseCoopers, "The Global State of Information Security Survey 2018," 2017.

[11] D. Woods and A. Simpson, "Policy measures and cyber insurance: a framework," *Journal of Cyber Policy,* vol. 2, no. 2, pp. 209–226.

[12] S. Romanosky, L. Ablon, A. Kuehn and T. Jones, "Content analysis of cyber insurance policies: how do carriers write policies and price cyber risk?," in *Workshop on Economics of Information Security*, 2017.

[13] B. Nieuwesteeg, L. Visscher and B. de Waard, "The law and economics of cyber insurance contracts: a case study," *European Review of Private Law,* vol. 26, no. 3, pp. 371–420, 2018.

[14] The CYBECO Consortium, "D7.1 – CYBECO Policy Recommendations," 2019.

[15] D. Banks, J. Rıos and D. Rıos Insua, Adversarial Risk Analysis, Francis and Taylor, 2015.

[16] R. D. Shachter, "Evaluating Influence Diagrams," *Operations Research,* vol. 34, no. 6, pp. 871–882, 1986.

[17] The CYBECO Consortium, "D3.1 – Modelling framework for cyber risk," 2018.

[18] The CYBECO Consortium, "D4.1 – Cyber-Insurance Use-Cases and Scenarios," 2018.

6

Cyber-Threat Intelligence from European-wide Sensor Network in SISSDEN

Edgardo Montes de Oca[1], Jart Armin[2] and Angelo Consoli[3]

[1]Montimage Eurl, 39 rue Bobillot, Paris, France
[2]CyberDefcon BV, Herengracht 282, 1016 BX Amsterdam, the Netherlands
[3]Eclexys Sagl, Via Dell Inglese 6, Riva San Vitale, Switzerland
E-mail: edgardo.montesdeoca@montimage.com; jart@cyberdefcon.com; angelo.consoli@eclexys.com

SISSDEN is a project aimed at improving the cyber security posture of EU entities and end users through development of situational awareness and sharing of actionable information. It builds on the experience of Shadowserver, a non-profit organization well known in the security community for its efforts in mitigation of botnet and malware propagation, free of charge victim notification services, and close collaboration with Law Enforcement Agencies (LEAs), national CERTs, and network providers. The core of SISSDEN is a worldwide sensor network which is deployed and operated by the project consortium. This passive threat data collection mechanism is complemented by behavioural analysis of malware and multiple external data sources. Actionable information produced by SISSDEN provides no-cost victim notification and remediation via organizations such as CERTs, ISPs, hosting providers and LEAs such as EC3. It will benefit SMEs and citizens which do not have the capability to resist threats alone, allowing them to participate in this global effort, and profit from the improved analysis and exchange of security intelligence, to effectively prevent and counter security breaches. The main goal of the project is the creation of multiple high-quality feeds of actionable security information that can be used for remediation purposes and for proactive tightening of computer defences. This is achieved through the development and deployment of a distributed sensor network based on state-of-the-art honeypot and darknet technologies, the creation of a high-throughput data processing centre, and provisioning of in-depth analytics, metrics and reference datasets of the collected data.

6.1 Introduction

The primary data collection mechanism at the heart of the SISSDEN project[1] is a sensor network of honeypots and darknets. The sensor network is composed of VPS provider hosted nodes and nodes donated to the project by third-parties acting as endpoints. These VPS nodes/endpoints are not the actual honeypots themselves. Instead, they act as layer 2 tunnels to the SISSDEN datacenter. Attack/scan traffic to the VPS nodes is sent via these tunnels to corresponding VMs which run the actual honeypots themselves. The honeypots in the datacenter then respond to the attacks/scans with the IP addresses from the VPS nodes.

This approach allows for easier management of the honeypots themselves – instead of having to remotely manage (and maintain) honeypots at the VPS provider locations, all can be centrally managed in one datacenter instead.

Each sensor endpoint has multiple IPv4 addresses – one for management, the others for tunnelling to the real honeypots.

As of 14th of January 2019, SISSDEN has 226 operational nodes running, spread across 58 countries. A total of 953 IP address from 112 ASNs are used, covering 375/24 networks.

The following world map (Figure 6.1) shows the current snapshot of operational sensor IPs:

Nine different honeypot types are currently deployed. These are focused on observing different forms of attacks against SSH/telnet services, general or specialised web services, remote management protocols, databases, mail relays, ICS devices, etc, including exploits, scans, brute force attempts. Information about these attacks is disseminated to 95+ National CSIRTs and 4200+ network owners via Shadowserver's free daily remediation feeds. These are marked with source 'SISSDEN'. One can subscribe to SISSDEN feeds via the SISSDEN Customer Portal (https://portal.sissden.eu).

To capitalise on the tools and knowhow from the H2020 SISSDEN project and assure the sustainability of the results, innovative real-time Cyber Threat

---

[1]SISSDEN (Secure Information Sharing Sensor Delivery event Network) is an H2020 project. See https://cordis.europa.eu/project/rcn/202679_en.html and https://sissden.eu/ for more information.

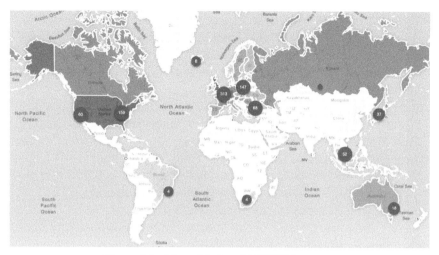

Figure 6.1 Map of deployed SISSDEN sensors.

Intelligence data for timely threat detection and prevention will be provided by a new start-up company called SISSDEN BV (https://sissden.com), launched by three SME partners (CyberDefcon, UK/The Netherlands, Montimage, France, and Eclexys, Switzerland).

6.2 SISSDEN Objectives and Results

6.2.1 Main SISSDEN Objectives

The main objectives of the SISSDEN project are:

- Create a large distributed sensor network. Over 100 passive sensors based on current and beyond state-of-the-art honeypot and darknet technologies are deployed in multiple organisations, including all 28 EU member states and 6 candidate countries, and are being used to observe malicious activities on an unprecedented scale, without intercepting any legitimate traffic.
- Advancements in attack detection. New types of honeypots, darknets and probes are deployed to detect, analyse and alert on types of attacks not widely detected today, such as reflective DDoS amplification or attacks against Internet of Things (IoT) devices, which are expected to increase significantly in the coming years as a range of new network-centric technologies are embraced by consumers and SMEs globally.

- Advancements in malware analysis and botnet tracking. The large sensor network is augmented by an innovative new generation of enhanced sandbox technologies designed for long running monitoring of malware specimen execution and behavioural clustering, to provide even more information on current threats.
- Improving the fight against botnets. Sensor and sandbox data collected is used for detailed studies of botnet infrastructures. Long-term observation of multiple families of current botnets will support anti-botnet research and law enforcement activities. Output will closely align with the existing European anti-botnet and anti-cybercrime strategies, as well as providing support to proven strong LEA partnerships, such as with Europol's European Cybercrime Center (EC3).
- Collect, store, analyse and reliably process Internet scale security data sets. The inherent challenges of building and continuously operating reliable data collection, storage, exchange, analysis and reporting systems at high volumes is solved by multiple innovations in sensor and backend packaging, deployment, integration and data searching, based on SISSDEN's consortium's extensive experience with "big data" approaches, high volume transactional and non-relational data systems.
- Share high-quality actionable information on a large scale. SISSDEN produces large amounts of intelligence on current threats and all of it is being shared with stakeholders and the larger community, at no cost to them, for the purposes of remediation or for early warning. The project distributes high-quality data feeds to the majority of the National CERTs in Europe, as well as worldwide, along with Law Enforcement Agencies, Internet providers, network owners and other vetted organisations fighting to defend their networks, SME customers, EU citizens and Internet users against continuous attacks.
- Provide objective situational awareness through metrics. Access to huge amounts of high-quality data on cyber threats: primarily obtained by the sensor network, but also contributed by the members of the SISSDEN consortium, provides metrics that offer objective, non-vendor biased overview of the threat landscape in the EU and individual member states.
- Create and publish a large scale curated reference data set. A significant subset of the data produced by SISSDEN is being made available to vetted researchers and Academia, addressing the clear and urgent need for large scale, high quality, and recent security datasets in order to improve or test defensive solutions.

6.2.2 Technical Architecture

Figure 6.2 below provides a simplified view of the SISSDEN technical architecture.

Components located at the EU datacentre include the Frontend Servers, Backend Servers and Utility Server pictured on the diagram. The sensor network consists of remote VPS Provider end points located at various VPS hosting providers (i.e. outside the EU datacentre), configured as transparent network tunnel endpoints forwarding traffic to the EU datacentre. SISSDEN collects attack data, such as network scans, spam email, malware binaries, brute force attacks, interactive attacker logins, etc.

6.2.2.1 Remote endpoint sensors (VPS)

Each remote endpoint sensor contains only the minimum amount of configuration and management capabilities required to securely participate as one end of a transparent network tunnel. They are configured to act as a long virtual Ethernet cable between the VPS and SISSDEN's local data centre frontend. At the Frontend in the EU Datacenter, a tunnel server terminates each transparent layer 2 Ethernet tunnel and delivers the Ethernet frames to an isolated, dedicated Virtual Local Area Network (VLAN).

6.2.2.2 Frontend servers

Traffic from the remote sensor endpoints are received by multiple types of honeypot systems, implemented as VMs, running on the EU Datacentre Frontend. Each honeypot VM emulates one or more potential vulnerabilities and collect data about attacks observed against those vulnerabilities. The honeypots have a standard configuration and standard data collection formats

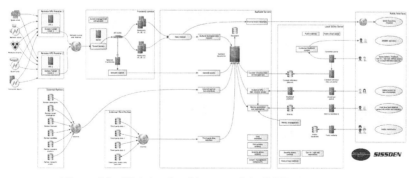

Figure 6.2   High-level architecture of the SISSDEN network.

enabled. Their data collection capabilities are complemented by network packet capture components (using solutions such as MMT and Snort) running on separate VM instances that listen to all traffic coming to them. SISSDEN system management components centrally manage all VM configuration, orchestration and operations.

Honeypot data and data from the network capture components are being ingested into the Backend datastores located at the Backend Servers at the EU Datacentre.

Tools like MMT and Snort are used to capture and analyse the network traffic. Snort allows identifying attacks using known attack signatures. MMT (Montimage's monitoring framework), adapted for SISSDEN, allows characterising malicious behaviour that corresponds to both known and unknown attacks. This information, referred to as CTI, is used by this monitoring framework for automating the real-time prevention and mitigation of attacks to an organisation (large or small) before they reach their network.

6.2.2.3 External partner and third-party systems

The data collected by the SISSDEN sensor network is supplemented by data from external systems operated by SISSDEN partners. These include separate honeypot networks, darknets, sandbox and malware analysis systems, threat intelligence platforms, etc. As with the sensor network, data from these systems is being ingested in various forms and stored in the Backend data stores.

To avoid unnecessary software development, SISSDEN makes use of and extends background partner systems, which aggregate data from multiple sources and provide a well-defined RESTful API for accessing normalized datasets.

6.2.2.4 Backend servers

Data from SISSDEN's various data collection systems is presented in multiple formats, such as live-streamed events, log files, PCAP files, and other file format data. Most of these data types are stored in their raw format in local data storage systems, at least for predetermined periods/repository size quotas, and some of the data types require parsing, normalization and ingesting into backend data indexes in support of free daily remediation report generation, high-value CTI, data analytics and ad-hoc querying.

6.2.2.5 External reporting system

One of the main purposes of the SISSDEN project is to collect Internet scale, timely security event data and make it available at no cost to vetted National CERTs, Network Owners and organizations who sign up for SISSDEN's free daily alerts.

The various sources of data collected by SISSDEN, such as honeypot and darknet data, malware analysis data, and botnet tracking information – as well as ingested external third party data sources – is being collected and stored locally in the SISSDEN backend. Each day, recipients who have voluntarily signed up for free reporting will receive by email multiple reports, covering different types of potentially malicious activity detected by SISSDEN on their nominated, verified IP/ASN/CIDR addresses.

On the other hand, SISSDEN BV will further provide real-time CTI, through a subscription service, to allow any organisation to block identifying cyber-attack campaigns before they reach their networks.

6.2.2.6 Utility server

Various analytics are being performed on the data collected by SISSDEN. An analytics platform is being extended, and hosted on the Utility Server. These analytics solutions provide additional insight into threats propagating in the Internet, pooling together partner resources dedicated to the project. In addition, metrics are being applied to the collected datasets to provide improved situational awareness. They can be used as a basis on which informed decisions can be made to mitigate threats. Curated reference datasets are also being made available to vetted researchers through the Utility Server. Interactions with the above are described in more details in this document and take place through the external interfaces illustrated in the diagram (with the exception of the analytics platform, which is only available to SISSDEN partners).

SISSDEN presents a number of systems to interact with the public and external partners. These include a Public website (mostly containing information about the project), email communication (reports), a Customer Portal, Metrics Dashboard, etc. Hosted on the Utility Server, these public facing systems include mechanisms to communicate with the consortium, sign up to request free of charge reports, gain access to the curated reference data set, provide customer feedback, and manage opt in/out and data privacy issues.

6.2.3 Concrete Examples

Two use cases, among many, have been selected to illustrate the real added-value of the CTI information that is provided.

6.2.3.1 Use Case 1: Targeted Cowrie attack that can be anticipated by the analysis of the traffic before it occurs

Targeted attacks are one of the emerging trends in cyber-security. Unlike conventional network scans and massive operations like spam and phishing, these attacks are generally answering the following criteria:

- They are focused on the assets of a single victim (private institution, government, critical infrastructure...) with objectives such as Data Exfiltration and Service Disruption.
- In the case of Data Exfiltration, it needs to be prepared and carried out after studying the infrastructure of the victim. The attackers will most probably put a lot of effort to hide their activity.
- In the case of Service Disruption, the attack is generally based on DDoS activity to disrupt the services and assets of the victim. This objective is normally achieved in a very short time (few minutes) and could be carried out repeatedly, thus generating an annoying service disruption, and consequently impact the victim's reputation. If the victim is a cyber-security company, the attack may take offline important security infrastructure (such as IDPS, honeypots and firewalls) and thus open the door to other attacks toward the protected zones (clients' assets, infrastructures...).

From a network traffic point of view, a targeted attack on honeypots looks like the curve shown in Figure 6.3. The spike shows when the targeted honeypot and/or its back-end system are hit. The graph shows the number of events registered in by the honeypot system which led to a 2-hour downtime of the honeypot system.

One can see the "normal traffic noise" before the attack and after the system has recovered.

Service suppliers (e.g., hospitals, media, power plants, control systems) cannot afford a 2-hour downtime. This class of attacks are able to disrupt the majority of infrastructures on the market. This has led the SISSDEN BV team to develop a DDoS resilient honeypot that will detect but not suffer from these attacks and therefore offer customers an improved security and uninterrupted threat analysis/monitoring.

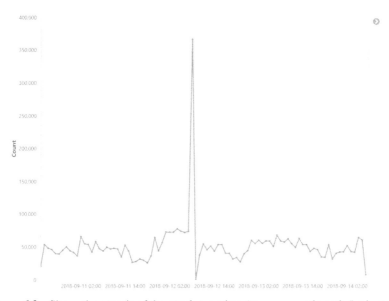

Figure 6.3 Shows the genesis of the attack over time (measures made each 5 minutes).

Furthermore, with the information that can be provided, the customers can prevent that their services do not go down in their own networks. For this, they can redirect or drop all the ingress traffic coming from the sources of this spike (using the IP addresses) and, if the attack starts occurring, set up another path for the egress traffic.

6.2.3.2 Use Case 2: Understanding the numbers – metrics

SISSDEN delivers realistic up-to-date metrics data and dashboards from its own sources that are compared and complemented with collated sources. SISSDEN categories are based on digital epidemiology and evidence-based practices as modelled from prior knowledge and research gained from other H2020 EU Projects: SAINT[2] and CyberROAD[3]. SISSDEN provided data can be used to make more informed decisions and improve security outcomes for clients. For instance, CTI data from SISSDEN and related sources found that in the first quarter of 2018 alone, the average enterprise faced:

---

[2] https://cordis.europa.eu/project/rcn/210229_en.html and https://project-saint.eu/
[3] https://cordis.europa.eu/project/rcn/188603_en.html and http://www.cyberroad-project.eu/

- 21.8% of all Website traffic that is due to bad bots (a 9.5% increase over the first quarter of 2017). For example, click fraud is a major threat especially for ISP's and enterprises, 1 out of 4 clicks are now fraudulent.
- 7,739 malware attacks (a 151% increase over the first quarter of 2017).
- 9,500 Botnet C&Cs (Command and Control servers) on 1,122 different networks (a 25% increase over the first quarter of 2017).
- 173 ransomware attacks (a 226% increase over the first quarter of 2017).
- 335 encrypted cyber-attacks (a 430% increase over the first quarter of 2017).
- 963 phishing attacks (a 15% year-over-year increase).
- 554 zero-day attacks (a 14% increase over 2017).
- 5,418,909,703 (5.4 billion) Web-based user accounts that have been compromised by 310 known or reported data breaches (a 40% increase over the first quarter of 2017).
- 40% of business and government networks in US and Europe shown evidence of DNS tunnelling.
- 75% of application DDoS, like HTTP-flooding, was in fact automated threats to Web applications mistakenly reported as DDoS.
- 73% of cyber-attacks focused on the cloud were directed at Web applications.
- 755 of 62,167 of the ASNs (autonomous systems) in routing system (1%) account for hosting, routing and trafficking 85% of all malicious activity.
- 13,935 total incidents are either route hijacks or outages. Over 10% of all ASNs were affected. 3,106 ASNs were a victim of at least one routing incident. 1,546 networks caused at least one incident in 2017 and already up by 20% in 2018.
- 90% of enterprises feel vulnerable to insider attacks, of which 47% are insiders wilfully causing harm and 51% are from insiders by accident; compromised credentials, negligence etc.

Ultimately analysing this type of metrics data by attack type, origin and region helps enterprises understand how cyber-attack trends are evolving. SISSDEN BV innovative AI approaches help in the timely prevention of these threats, remove false positives, help improve budget/resources prioritisation, and improve awareness with open source feeds.

6.3 Conclusion

Many security-oriented tools and services exist that provide or use CTI for the prevention, detection and response to threats. CTI is integrated natively into security products (i.e., appliances and software tools) or provided as a service for organisations' response teams. Among those that offer state-of-the-art Threat Intelligence solutions and services we have, for instance [2]: Anomali, ThreatConnect, ThreatQuotient, LookingGlass and EclecticIQ.

With respect to these offers, the SISSDEN project provides free feeds derived from its wide network of honeypots and darknets; and, the start-up, SISSDEN BV, provides original real-time actionable feeds complemented with information from other sources, that are not provided by these companies since they mainly rely on the existing open data that is analysed offline.

The innovation with respect to state-of-the-art market solutions provided by SISSDEN concerns the following:

- Ease of use and comprehensive threat indicators: SISSDEN relies on open standards (e.g., STIX/TAXII) and provides malicious-only IP addresses, subnets, URLs, threat ontology and ASNs.
- Trust in provided intelligence and accuracy: SISSDEN intelligence comes from malicious honeypot and darknet activity that contains no false positives.

The SISSDEN BV start-up further provides:

- Timely and Real Time: SISSDEN BV delivers CTI in real time (less than 1 minute) for effective blocking of attacks before they occur.
- CTI is correlated with information from other sources and using Deep Data and Artificial Intelligence-based analysis, increasing its value and extent.
- Removing complexity: SISSDEN BV allows for efficient use of security resources and provides shared threat intelligence and automated response.
- Modular and scalable: SISSDEN BV can serve different categories of customers: SMEs without security expertise or solutions, medium and large enterprises with their own solutions and security teams...

Acknowledgements

This work is performed within the SISSDEN Project with the support from the H2020 Programme of the European Commission, under Grant Agreement No 700176. It has been carried out by the partners involved in the project:

- Naukowa I Akademicka Siec Komputerowa, Poland
- Montimage EURL, France
- CyberDefcon LTD, United Kingdom and The Netherlands
- Universitaet des Saarlnades, Germany
- Deutsche Telekom AG, Germany
- Eclexys SAGL, Switzerland
- Poste Italiane – Societa per Azioni, Italy
- Stichting The Shadowserver Foundation Europe, The Netherlands.

References

[1] Bachar Wehbi, Edgardo Montes de Oca, Michel Bourdellès: Events-Based Security Monitoring Using MMT Tool. ICST 2012: 860–863

[2] Craug Lawson, Khushbu Pratap; "Market Guide for Security Threat Intelligence Products and Services" published 20 July 2017 by Gartner.

7

CIPSEC-Enhancing Critical Infrastructure Protection with Innovative Security Framework

**Antonio Álvarez[1], Rubén Trapero[1], Denis Guilhot[2],
Ignasi García-Mila[2], Francisco Hernandez[2], Eva Marín-Tordera[3],
Jordi Forne[3], Xavi Masip-Bruin[3], Neeraj Suri[4], Markus Heinrich[4],
Stefan Katzenbeisser[4], Manos Athanatos[5], Sotiris Ioannidis[5],
Leonidas Kallipolitis[6], Ilias Spais[6], Apostolos Fournaris[7]
and Konstantinos Lampropoulos[7]**

[1]ATOS SPAIN, Spain
[2]WORLDSENSING Limited, Spain
[3]Universitat Politècnica de Catalunya, Spain
[4]Technische Universität Darmstadt, Germany
[5]Foundation for Research and Technology – Hellas, Greece
[6]AEGIS IT RESEARCH LTD, United Kingdom
[7]University of Patras, Greece
E-mail: antonio.alvarez@atos.net; ruben.trapero@atos.net;
dguilhot@worldsensing.com; igarciamila@worldsensing.com;
fhernandez@worldsensing.com; eva@ac.upc.edu; jforne@entel.upc.edu;
xmasip@ac.upc.edu; suri@cs.tu-darmstadt.de;
heinrich@seceng.informatik.tu-darmstadt.de;
katzenbeisser@seceng.informatik.tu-darmstadt.de;
athanat@ics.forth.gr; sotiris@ics.forth.gr; lkallipo@aegisresearch.eu;
hspais@aegisresearch.eu; apofour@ece.upatras.gr; klamprop@ece.upatras.gr

In the recent years, the majority of the world's Critical Infrastructures (CIs) have evolved to be more flexible, cost efficient and able to offer better services and conditions for business growth. Through this evolution, CIs and companies offering CI services had to adopt many of the recent advances of the Information and Communication Technologies (ICT) field. This rapid adaptation however, was performed without thorough evaluation of its impact on CIs' security. It resulted into leaving CIs vulnerable to a new set of threats and vulnerabilities that impose high levels of risk to the public safety, economy and welfare of the population. To this extend, the main approach for protecting CIs includes handling them as comprehensive entities and offer a complete solution for their overall infrastructures and ICT systems (IT&OT departments). However, complete CI security solutions exist, in the form of individual products from IT security companies. These products, integrate only in-house designed and developed tools/solutions, thus offering a limited range of technical solutions.

The main aim of CIPSEC is to create a unified security framework that orchestrates state-of-the-art heterogeneous security products to offer high levels of protection in IT (information technology) and OT (operational technology) departments of CIs, also offering a complete security ecosystem of additional services. These services include vulnerability tests and recommendations, key personnel training courses, public-private partnerships (PPPs), forensics analysis, standardization activities and analysis against cascading effects.

7.1 Introduction

7.1.1 Motivation and Background

Critical infrastructures (CIs) are defined as systems and assets either physical or virtual, extremely vital to a state. The incapacitation or destruction of such infrastructures would have a debilitating impact on security, economy, national safety or public health, loss of life or adversely affect the national morale or any combination of these matters. These infrastructures affect all aspects of daily use including oil and gas, water, electricity, telecommunications, transport, health, environment, government services, agriculture, finance and banking, aviation and other systems that, at the basis of their services, are essential to state security, prosperity of the state, social welfare and more.

In the recent years, the majority of the world's CIs has unstoppably evolved to be more flexible, cost efficient and able to offer better services and business opportunities for existing but also new initiatives. CIs and companies offering CI services had to adopt many of the recent advances of the Information and Communication Technologies (ICT) field, thus incorporating the use of sophisticated devices with improved networking capabilities. In fact, the use of Internet enables a distributed operation of facilities, an optimized sharing and balance of resources through network elements, eases the prompt notification and reaction in case of emergency scenarios. In parallel, physical devices like sensors, actuators, engines and others become more and more intelligent thanks to the recent Internet of Things paradigm. In most cases, however, these advances have been performed, without security in mind. Apart from the security risks imposed by the new connections to the Internet, there are also additional risks due to IT/OT software vulnerabilities. The result was to leave CIs vulnerable to a whole new set of threats and attacks that impose high levels of risk to the public safety, economy and welfare of the population. One example of these vulnerabilities is the WannaCry incident, produced by a ransomware attack [1], in 2017 that affected more than 200,000 Windows systems, including CIs such as six UK hospitals of the Britain's National Health Service (NHS). Other data, show that the number of incidents in the power supply systems sector has increased from 39 in 2010 to 290 in 2016 [2], including the cyberattacks to the Ukrainian power supply plant in 2015 and 2016.

This data and considering that the borders between OT and IT sides of CIs have progressively blurred, show that CIs have become more exposed to the public through Internet and therefore within reach of cyber criminals. The landscape of possible attacks against critical infrastructures has widen a lot and is still evolving at a very quick pace. Some examples include cross-site scripting attacks, code injections of any kind, with SQL injection being one of the most popular ones, malicious files uploads, virus installation via USB, ports scan & intense network scans, binary trojans, denial of Service (DoS), email propagation of malicious code, spoofing, botnets or worms, to name some. Also we cannot neglect that personal information belonging to CI users may be compromised, jeopardizing more than just their privacy. To respond to this, the CIPSEC project has developed the CIPSEC framework for critical infrastructure protection, which is presented within the next sections.

7.1.2 CIPSEC Challenges

Critical infrastructures (CIs) consist of several different, heterogeneous subsystems and need holistic solutions and services to provide coverage against a broad range of cybersecurity attacks. The main objective of the

CIPSEC project is to create a unified security framework that orchestrates state-of-the-art heterogeneous, diverse, security products and offers high levels of cybersecurity protection in IT and OT CI environments. CIPSEC Framework should be able to collect and process security-related data (logs, reports, events), to generate anomaly based security alerts for events that can affect CI's health and can have a series of cascading effects on other CI systems. The developed framework should be very flexible and adaptive to any CI. Additionally, it should cause minimum interference to the CI's normal functionality and should be able to upgrade its components, when an update is available in a secure and easy manner.

Beyond that, CIPSEC aims to provide a series of services to support the CIs in attaining a high cybersecurity level. Specifically, CIPSEC provides CIs' systems vulnerability tests and recommendations including studies for cascading effects, promotes information sharing and describes good security policies that need to be followed by the CI administration and personnel. The CIPSEC framework incorporates a training service that will assist the CI's personnel how to use the proposed framework, as well as basic cybersecurity principles to be followed in the CI routine. Finally, we also introduce an updating and patching mechanism to keep the framework always updated and secure against the latest cyber attacks.

To prove the effectiveness and efficiency of the CIPSEC framework and to evaluate the security level of the solution, we have installed our solution in real conditions, inside three pilot infrastructures belonging to the transportation, health and environment monitoring sectors respectively. Using the output and knowledge derived from the three-pilot experimentation, we aim on communicating the CIPSEC results to standardization bodies and influence emerging standards on CI security primarily in transportation, health and environmental monitoring and in other CI domains (like smart grid or industrial control). Finally, the CIPSEC ultimate objective is to create a framework solution that can enhance the current cybersecurity market and has a positive impact on the CI cybersecurity ecosystem. CIPSEC's goal is to provide a solution that is market ready, innovative and well beyond the relevant market competition, thus offering interesting business opportunities and exploitation results.

The rest of this chapter is organized as follows. Section 2 presents the innovations of the project. Section 3 describes the CIPSEC framework, including the proposed architecture. Section 4 shows how the proposed solution is applied to the three different pilots. Section 5 addresses dissemination and exploitation. Finally, Section 6 concludes the chapter.

7.2 Project Innovations

Each individual solution introduced must successfully match all the requirements of the Critical Infrastructure Security domain and be fully compatible with the overall CIPSEC framework technical and market goals. Moreover, it must be viewed as a commercial solution and, as such, target individually and through the CIPSEC framework a specific part of the relevant market. Thus, all the CIPSEC security products/solutions are designed with strong innovation in-mind, to better achieve strong technical and market benefits. The CIPSEC anomaly detection reasoner, namely the ATOS XL-SIEM product. IT can integrate inputs from many heterogeneous observable indicators of cyber-attacks without any compromising its reliability. Also, the XL-SIEM system can support even legacy monitoring equipment (typically found in long-lifetime critical infrastructures). XL-SIEM introduces intelligence into the traditional correlation ecosystem that exists today, providing information and visibility of the cybersecurity events produced inside organizations in real time. It consists of a real-time distributed and modular infrastructure, that adapts to the specific needs of each organization. Sensors of the CIPSEC anomaly detection reasoner are innovative themselves. For instance, the Bitdefender antimalware solution can provide proactive detection for previously unseen malwares with an uncharted behaviour. In a way, the antimalware solution is capable of detecting anomalies in the system's behaviour even if they are unknown to it through the introduction of new technologies like deep packet inspection and machine learning techniques. Innovative honeypot solutions are integrated and combined to capture and analyse a broad range of attacks. They can analyse IT and OT infrastructure traffic and create replicas of real IT and OT services. It also includes peripheral security solutions like rootkit hunters and SSH attack detectors. Moreover, CIPSEC solution incorporate a series of honeypots that are able to detect attack attempts prior to happening and divert attacks from the production systems to them. The honeypot solutions consist of a DDoS amplification honeypot, a low interaction honeypot and an ICS/SCADA honeypot. The CIPSEC framework, innovated by introducing apart from software-based solutions also hardware security solutions. Denial of Service attacks on the physical layer of broad wireless band can also be detected in an innovative way by DoSSensing, that operates as an external element sentinel to specifically detect Jamming attacks to any band(s) in which the wireless sensors, industrial IoT elements, and even computers connect to the Critical Infrastructure network. Empelor's innovative programmable, flexible and diverse card reader solution can be adapted to any critical infrastructure environment at hand and that offers multi factor authentication. The framework also includes a Hardware Security Module solution that is directly connected to CI host devices and acts as a trusted environment for security/cryptography related operations and secure storage. This solution is extremely fast since computation intensive cryptography operations are accelerated by hardware means and thus fits well to the critical, real-time nature of many CI systems. Another important feature that the CIPSEC framework offers is the ability to visualize forensics events. By implementing and installing in the CI system Critical Infrastructure Performance Indicators (CIPIs), we are able to collect, analyse and visualize forensics measurements. Thus, we are able to innovate by providing advanced, intuitive and detailed data visualizations to active (real time) cyber/digital forensics analysis where data from heterogeneous sources are aggregated, combined and presented in a intuitive manner. Finally, the CIPSEC framework can handle private data by including and applying anonymization methodologies through a relevant tool wherever CI system needs it. The tool is based on innovative research on micro-aggregation methods and fast computational responses for anonymizing data.

Apart from innovation from individual components of the CIPSEC Framework, the integration process of those heterogeneous components into

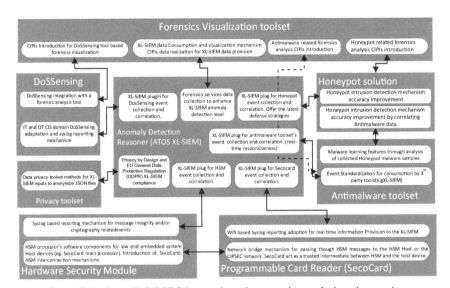

Figure 7.1   Overall CIPSEC innovations due to various solutions integration.

a unified, fully functional architecture has introduced several innovative aspects. Those aspects include the acquisition, exchange and management of security related information (events, logs, alerts) existing within the CIPSEC framework. Thus, every component of the framework has introduced some data exchange mechanism or feature to its architecture to be compliant, integrated-ready to the overall CIPSEC architecture. In the following figure (Figure 7.1), all such mechanisms/features are presented and described in brief.

7.3 CIPSEC Framework

This section is structured as follows. Firstly, the CIPSEC reference architecture for critical infrastructure protection is introduced. This architecture considers the basic data flow which takes place in all critical infrastructures. Once the architecture is defined, the functional components are detailed differentiating between core components and data collectors. Finally, the methodology followed for the integration of the components into a unified framework is explained.

7.3.1 CIPSEC Architecture

As presented in [3], the CIPSEC reference architecture, is proposed at data flow level, that is infrastructure agnostic and establishes a general framework for protection applicable to any critical infrastructure, regardless of the vertical (i.e. activity sector) it belongs to or the resources managed. In this sense, the architecture is flexible and adaptable. The architecture is based on the security data lifecycle existing in critical infrastructures and shared among their components.

The data lifecycle considers three different stages: data acquisition, dissemination and consumption.

- Data acquisition is the Critical-Infrastructure-specific devices that are used to acquire the data as specifically applicable to the Critical Infrastructures' industrial control, meaning that the control of a single process or machine is not interrelated directly to another process or machine. For example, hospital ventilators do not interact directly with syringe plumps. However, in Industry 4.0 IoT scenarios, all facilities tend to communicate with each other, increasing the security management complexity in such interoperable scenarios. At this stage of the lifecycle, communications are usually not done through public/open or documented protocols but through a proprietary protocol documented at the discretion of the manufacturer. And in most cases, monitoring is done through a client-server protocol. Some OT devices add to their own communication protocols the possibility of communicating using standard protocols such as Modbus, DNP3 or OPC UA. Not only OT field devices are involved in this stage, but also others like PLCs, robots or HMIs. Data transmitted are signals or data sequences used with different purposes, like for instance monitoring status.

- Data dissemination considers a set of networks, equipment and communication protocols that perform real-time monitoring of industrial processes and complex tasks that use the information obtained in the data acquisition phase. Data dissemination is also about communicating with actuators/controller devices to transmit to sensors appropriate orders that can control the process automatically by means of specialized software. In the data dissemination phase, the communication is facilitated through specific protocols between OT devices and OT controllers. Data dissemination is also about integrating and centralizing all signals generated by a given process. The data are monitored, controlled and managed in real-time. In data dissemination SCADA systems, OPC servers, activity monitoring systems or Historian servers are some examples of elements involved, while the information disseminated is related to process variables, consumed resources, downtimes or device status, for instance.

- Data consumption is associated to the concept of Industrial Business Intelligence (IBI), which in turn is defined as the set of tools, applications, technologies, solutions and processes that allow different users to process the collated information for decision making-purposes by using the sensory and behavioural data as collected from network infrastructures. This information is the result of a process which starts by extracting the information from different data sources. Then, there is a transformation process consisting of contextualizing the raw data obtained from such different data sources. Finally, the loading process consists of storing all the information already contextualized in some centralized data storage point. Several tools will take care of exploiting the information once it is available in the data storage point, these tools are focused on offering the user several KPIs that allow to make informed decisions.

On the basis of this data lifecycle, the next challenge addresses the security aspects relevant to the critical infrastructure. CIPSEC proposes to integrate the security data lifecycle around the critical infrastructure data lifecycle to decouple both processes and avoid conflicts. The approach used is similar, using exactly the same stages: acquisition, dissemination and consumption.

For data acquisition, CIPSEC considers a wide range of data sources such as Host Intrusion Detection Systems (HIDS), Network Intrusion Detection Systems (NIDS), data from other systems that coexist together in the same security ecosystem, log files, monitoring status information, reports and human knowledge. It is relevant to highlight the utility and variety of information that can be obtained from the logs. To provide some examples, these logs can contain information about firewalls, antivirus/antimalware, real-time activity monitoring, intrusion detection sensors or disturbances in wireless signal. The CIPSEC Framework uses a combination of detailed event logs, collected from heterogeneous security solutions, used to provide a complete audit trail covering data acquisition to data delivery.

Data dissemination addresses how the information is made available to different stakeholders and systems. The organization should disseminate security knowledge to stakeholders, especially about security incidents, and focus on establishing a dissemination plan to deliver critical knowledge, more specifically to get the right information, transform it in the right format, to the right people, and at the right time. Types of outputs from this stage include events, alarms, tokens, software updates and security data insights.

Data consumption corresponds to the highest level of security management, obtaining an overview of the cybersecurity posture at all levels (for example, information about threats or attacks affecting the infrastructure), assisting to make timely decisions about prevention or mitigation of existing or upcoming attacks. Data consumption is all about understanding the critical infrastructure security data and extracting security insights from them. Decision-making is undoubtedly the main driver of this security data lifecycle. The complexity of the critical infrastructure processes requires carrying out decision-making activities both at business and technical levels. Profiles to be involved in the process may be field service technicians, network managers, security analysts, computer forensics, system administrators, contingency plan designers or industrial engineers to name but a few.

Once the picture is clear with respect to the data lifecycle in critical infrastructures, both for operational and security data, and insisting on the fact that the two cycles are completely decoupled and unrelated, the foundations are established for the definition of the CIPSEC reference architecture. To produce this architecture, the set of requirements expressed by the three

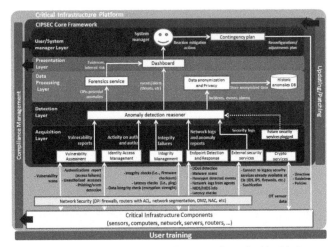

Figure 7.2   CIPSEC reference architecture for protection of critical infrastructures [4].

pilots of CIPSEC [5] were also considered, as well as the commonalities existing in critical infrastructures across different domains [6, 7] . Figure 7.2 shows this architecture, which is a very relevant result of the CIPSEC project. It is a layered architecture where the layers are established following the security data flow from the infrastructure to the user interface and back to the infrastructure to communicate the decisions made by the users (ideally made jointly by managers and technicians). This architecture is extensively explained in [3] and minor updates are reflected in [4].

The CIPSEC reference architecture is applied to the critical infrastructure itself, considering its operative components and the deployed network security. CIPSEC makes a leap forward protecting the whole perimeter of the critical infrastructure and therefore enhancing its security. The closest layer to the infrastructure is the acquisition layer. This element, consists of five main components: Vulnerability Assessment, Identity Access Management, Integrity Management, Endpoint Detection and Response, and Crypto Services. These components are able to obtain different inputs from both the critical infrastructure components and the network security elements. This layer also includes a block for future security services that can be plugged into the framework. On top of the acquisition layer, the detection layer is placed. This layer includes the Anomaly Detection Reasoner which receives aggregated information from the different acquisition layer blocks.

On top of the detection layer, the data processing layer includes the Data Anonymization and Privacy Tool, capable of anonymizing sensitive data coming from the critical infrastructure and eventually storing it in a historic anomalies database. The data processing layer also contains the forensics service, which receives critical infrastructure performance indicators and produces relevant information that can be used in a forensics analysis upon incident occurrence. The presentation layer is implemented as a dashboard which shows a summary of the main highlights concerning the security status of the critical infrastructure, and also offers the specific details provided by the user interfaces of the different components which are integrated in a harmonized way with a common look and feel. All the details about the dashboard are documented in [8]. The information in the dashboard can be used to decide on reaction mitigation actions and to produce a sound contingency plan with reconfigurations and adjustments to be applied to the infrastructure. With regards to this, CIPSEC provides a consulting service aiming at assisting the user to produce a complete contingency plan. Three more services are present in the architecture: the compliance management service, that is part of the contingency service and its goal is to show the level of compliance between the solutions provided by the CIPSEC Framework and the requirements of the respective critical infrastructure. Another service, applies updates and patches, in an automated manner, to the components of the framework when it is required. Finally, the framework also offers a training service aiming at improving the skills of the operators in charge of managing the security of the critical infrastructure.

The solutions provided by the partners put in place the functionalities required to enable the different blocks of the architecture presented above to play their role in the integrated framework. These products and services fit well into the architecture and allow to establish the settings for the instantiation of the architecture in different scenarios, starting by those of the three pilots (presented in Section 4).

7.3.1.1 CIPSEC core components

CIPSEC core components are in charge of making the most of the information obtained by the collectors presented in Section 7.1.2. Their role is different depending on the component in question. The XL-SIEM (ATOS) correlates and processes events across multiple layers, identifying anomalies, and is present in the Anomaly Detection Reasoner component. The anonymization tool (UPC) implements different data sanitization mechanisms, including suppression, generalization and pseudonymization, to protect sensitive personal information. It is present in the Data Anonymization and Privacy component and makes it possible to share cybersecurity data among different critical infrastructure stakeholders without jeopardizing the privacy of the users. The Forensics Visualization Tool (AEGIS) provides intuitive and detailed visualizations to enable cyber/digital forensics analysis. It is present in the Forensics Service component. Finally, the dashboard is a vital core component, whose objective is to provide a unified, harmonized, and consistent application, where the user/administrator of the infrastructure is able to i) check for the current status; ii) easily access to all tools and services provided by the CIPSEC Framework and iii) be warned about current or future threats in the system

7.3.1.2 CIPSEC collectors

CIPSEC combines information produced by the different products playing a role within the acquisition layer of the framework. They monitor OT systems and collect raw security data from multiple sources and functionalities and provide monitoring and anomaly detection for the complete critical infrastructure. The collectors are the following:

The Forensics Agents (AEGIS) are a set of plugins/tools deployed in the critical infrastructure and properly configured to log information that is relevant to the hosting critical infrastructure and is used by the Forensics Service. The Network Intrusion Detection System (ATOS) sensor is similar to a sniffer since it monitors all network traffic searching for any kind of intrusion. It implements an attack detection and port scanning engine that allows registering, alerting and responding to any anomaly previously defined as patterns. The Gravity Zone Antimalware Solution (Bitdefender) detects malware, phishing, application control violation or data loss, among others. Honeypots brought by FORTH monitor the critical infrastructure network and produce insightful results for the anomaly detection and prevention component. Used Honeypots are Dionaea, Kippo, Conpot and a custom DDoS honeypot based on the detection of amplification attacks. The DoSSensing Jamming Detector by World sensing monitors the whole wireless spectrum to detect anomalies derived from a Denial of Service attack in real-time. All the aforementioned solutions are present in the Endpoint Detection and Response component. The Hardware Security Module developed by the University of Patras is a synchronous Secure System on Chip (SoC) device implemented on FPGA technology. It is a trusted device offering cryptography, secure storage and message integrity services. It is present in the Crypto Services and Integrity Management components. Secocard (Empelor) is a security enhanced single board embedded microcontroller and is present in the Identity Access Management, Integrity Management and Crypto Services component.

7.4 CIPSEC Integration

CIPSEC is a challenging project in terms of integration. The goal is to obtain an orchestrated solution which offers a general yet comprehensive approach to protect critical infrastructure against cyber threats. A clear roadmap for component integration was designed by the Consortium to produce the solution that makes the most of the features of the components brought by the different project partners. A thorough study of each product was carried out to understand the kind of information that can be obtained from such product. An important aspect to analyze was how this information was provided. In the specific case of the products playing the role of collectors (see Section 7.1.2), they produce logs containing the relevant information to consume. These logs have different formats according to the kind of event to communicate and also depending on the product in question. The CIPSEC architecture proposes the Anomaly Detection Reasoner (with the ATOS XL-SIEM playing this role) as the orchestrating element that integrates the logs from a wide range of collectors. To do so, several plugins were developed to adapt the different formats to the one understandable by the XL-SIEM. As the plugins were available and therefore the information coming from the different collectors was translated into the common format, the partners researched on how to combine events coming from different products to produce more complex events and eventually alarms with insightful messages demanding actions and clear responses from the user. Regarding the core components (Section 7.1.1), it is important to highlight the approach used for the dashboard (see Figure 7.3), which embeds views from the different products under a common look and feel, offering a harmonized user interface for the different CIPSEC user profiles.

All the development and tests were carried out on a distributed testbed where the different components were located in public IPs within each partner's local network, resulting on a testbed distributed in countries like Spain, Greece, United Kingdom, Romania and Switzerland. This led to a distributed prototype ready to be deployed in the three pilots. A deployment plan was designed for each pilot. The prototype is flexible enough to allow the user to choose components off-the-shelf according to his specific needs. The three deployments were carried out in Darmstadt, Germany, for the railway pilot; in Barcelona, Spain, for the health pilot; and in Torino, Italy for the environmental pilot. More details about the pilots are provided in Section 7.4. The approach for these pilots is hybrid, with most components deployed in the cloud except for those that necessarily need to be on premises,

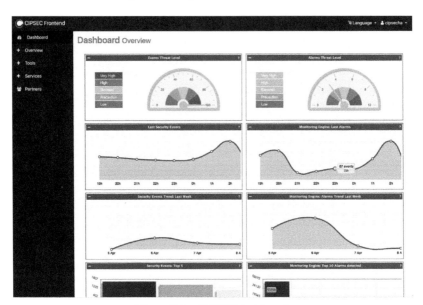

Figure 7.3   CIPSEC dashboard.

like the security data collectors. The prototype contains extra features like the presence of tools to produce attacks and to test its performance or a set of virtual assets emulating industrial networks. In some cases, critical infrastructures demand a completely on premise deployment, taking place in an off-line environment, without any connectivity to the Internet, as they work in isolation, therefore Internet connection is not an option for them. Based on this, a second prototype was created with the purpose of demonstrating CIPSEC even without internet connection. This prototype is composed of two physical machines that contain the CIPSEC solution in the form of several virtual machines. The deployment plan is to place all the VMs in the same local subnet where CI systems resides. Additionally, CIPSEC members use this prototype for demos in different events. All the details about the different integration environments and pilot deployments can be found in [8].

7.5 CIPSEC Pilots

The security framework for Critical Infrastructures (CI) proposed by CIPSEC has been designed, integrated, deployed and tested in 3 different pilot domains. Health sector- represented by Hospital Clinic de Barcelona

(HCB [9]), Transportation sector– represented by the German Railway infrastructure (Deutche Bahn [10]) and Environmental monitoring sector – represented by The Regional System of Detection of Air Quality AQDRS managed by CSI – Piemonte (CSI, [11]). For all the pilots of each domain, CIPSEC followed an analytical process of defining their characteristics, eliciting the requirements in terms of fitting solutions, analyzing the involved security and privacy aspects and finally extracting the system requirements, as described in the previous chapter. The integration and testing of the proposed solution is described in the following sections.

7.5.1 Integration of the Solution in the Pilots

Integration of security technologies in a CI is affected by a set of limiting factors that were faced by the CIPSEC team during the integration phase into the OT and IT systems of the pilots. Some indicative examples are communication infrastructure not following proper security guidelines (e.g. lack of firewalls), proprietary communication protocols, dedicated software that can't be managed by standard security tools, unattended physical locations of equipment, highly regulated environments, limitation/lack of resources, requirement for real time readiness and difficulty in applying patches and updates to existing working systems. To overcome these limitations, CIPSEC has developed a compliance management service (CMS) which shows the level of compliance between the solutions for cybersecurity that the CIPSEC framework provides and requirements of the CI stemming from various sources, such as expert knowledge, domain standards, industrial standards, or legislation. This process is performed by matching the CIPSEC Profile and the CI Profile so as to define the CIPSEC solutions that can be applied to the CI and therefore proceed with their deployment. So, the main objectives and the environment to test how CIPSEC can fulfil them were defined for all pilots.

The main objective of railway transportation is safe operation. Due to this the systems have to fulfil the requirements of several safety standards (EN 50126, EN 50128, EN 50129) and an admission by the national safety authority has to be granted. This also applies, if changes are made to the system which affect safety.

A typical control system in the railway domain consists of several subsystems:

- Safety-related components like interlocking, points, switches and axle counters

- Assisting systems like train number systems and automated driveway systems
- Data management systems as the MDM, the documentation system
- Diagnosis systems

The most relevant to CIPSEC components are the ones responsible for signaling, like interlocking systems. Due to the safety-relevance of the interlocking components and the required admission by the German national safety authority Eisenbahnbundesamt (EBA), DB established a test site in their OT testing facilities for testing the CIPSEC Framework. The environment consists of Operating Centers and operator workstations that simulate the normal operation of the system and therefore integration of CIPSEC components has been performed on a really close to real environment.

The Health pilot includes an abundance of in-hospital devices, many different networks with high low-latency constraints, controls at different levels and strong privacy requirements on the collected and processed data. HCB focused on the selection of the most representative IoT elements to be tested and the definition and construction of appropriate test sites. Due to the unavailability of these areas, the necessity to have them perfectly controlled (either from physical and remote accesses), the requirement to install the selected equipment inside as a local network but working separately from the central production servers of the data center and the lack of technical space dedicated to non-care uses inside the Hospital, HCB took action to:

- Adapt one existing test room dedicated to clinical emergency training to configure the test site 1 which includes medical equipment
- Adapt one existing office dedicated to new developments and technological trials to configure the test site 2 which includes IoT industrial equipment interacting with information provided by medical equipment
- Build from scratch a third room with the purpose of using it as test site 3 including generic IOT equipment

Having all these test sites available allowed CIPSEC to integrate all its components and design tests covering a plethora of usage scenarios for the hospital devices.

In the Environmental monitoring pilot, CSI is responsible for the monitoring network operated by ARPA Piemonte (Regional Agency for the Protection of the Environment of Piedmont region) which includes 56 monitoring stations and one Operations Center (OC) which receives the gathered environmental data. Protecting the stations and primarily the OC is the main objective of CSI. The pilot consists of five main functional areas:

- The air measurement equipment
- The PC Stations
- The OC Operations Centre Server for data acquisition
- The OC Operations Centre Databases
- The ARPA Enterprise Infrastructures

The CSI in agreement with ARPA prepared a testing environment. The virtualized environment is comprised of two parts: the monitoring station and the Operation Centre. Therefore, the CIPSEC components were integrated in this environment so as to test security threats regarding the normal operation of the stations, the uninterrupted communication with the OC and also other possible external cyber-attacks. It must be noted that all the aforementioned testing facilities included the deployment of new hardware and software components that allowed the creation of VLANs and the integration of the various CIPSEC components in networks local to the pilot CIs. All CIPSEC solution providers followed the integration guidelines described in Chapter 3 to successfully deploy their components to the test facilities and evaluate the CIPSEC prototype in all three pilot domains.

7.5.2 Testing the Proposed Solution in the Pilots

CIPSEC has followed a detailed testing methodology with regards to evaluation of performance and capabilities of the integrated platform. This methodology ("IEEE Standard for Software Test Documentation [12]") includes the definition, implementation execution and reporting of composite test scenarios that can prove the effectiveness of the CIPSEC platform in trial as well as real-world scenarios. The composite tests were defined for each one of the pilots and produced results covering many features of device resources and security requirements at the same time. Overall, 29 composite tests were executed in planned online and on-site sessions for all the pilots. [13] reports the execution results of these tests in detail, whereas recorded versions of the test execution are available for all the testing sessions.

Moreover, required equipment, the procedures and the people necessary to set up the CIPSEC tools were also recorded to identify problems and gain insights on possible deployment issues in real world deployments. The latter is also enhanced by the findings derived, after the tests were conducted. The main identified issues have to do with CIPSEC components requesting internet access, overall configuration of the framework being cumbersome and increased resource consumption by the components. To this end, the CIPSEC final prototype will offer a fully on-premise deployment that requires no internet connection to operate and an operational environmental that will allow tailor-made presentation of the infrastructure information that is of most importance to the CI managers.

7.6 Dissemination and Exploitation

7.6.1 Dissemination

Although finding solutions for protecting CIs is the main objective of the CIPSEC project, communicating and regularly showing the achieved progress will ensure the objectives are being accomplished. All CIPSEC results will be used to raise the citizens' awareness about CIPSEC solutions, paying special attention to target groups and the research community as a whole. In this sense, one of the first tasks in dissemination was to identify these possible target groups potentially interested in different aspects of the project. We identified seven target groups: Local Authorities, Policy Makers, Business people, Researchers, Associations, General Public and Media. A second task of the dissemination strategy was to create the approach and communication strategies to reach out to the identified stakeholders. Some of these communication activities include, the creation of a corporate identity, the maintenance and updating of a website, the production of promotional material, a monthly CIPSEC blog entry to disseminate the project ideas to a wide audience, the dissemination of daily information in CIPSEC social accounts (Twitter, LinkedIn and ResearchGate), the production of project videos and upload them in YouTube, and finally to produce scientific publications with research related to CIPSEC. During the life of the project we have already produced 23 blog entries, 10 videos in YouTube and we have 40 accepted papers.

7.6.2 Exploitation

The main strength of the CIPSEC framework is the integration and orchestration of heterogeneous solutions under one unique umbrella which is specifically designed to protect CIs. The pilots are an excellent showcase of the direct operational benefits for the Health, Transportation and Air Quality customer segments and the stakeholders' opinion will be extremely valuable to define a final market approach. It has been demonstrated that the targeted market is not necessarily aware not only of the solutions available, but sometimes of the actual pain points and issues it is facing. Technology evangelism is being performed and will be increased through the implementation of a free version of the CIPSEC framework which should be considered as a powerful demo of the capabilities of the premium version and provides a set of expressly selected functionalities. These will be strongly limited, and any additional customer support request will entail the regular applicable fees. The following remarks were taken into consideration:

- The free version must be representative of the full CIPSEC concept: stakeholder should get a quick idea of the premium version by simply interacting with the tool;
- Tools and services must be there: CIPSEC is not just the sum of different tools, but also services;
- The free version should be attractive and simple enough to gain the interest of heterogeneous stakeholders' groups.

In contrast, the premium version will merge the different solutions from all the partners, offering a full functionality on a business-based approach. The Consortium rules out the emerging of a joint company or a similar stable structure but instead it considers the establishment of a framework of collaboration to commercialise the collaborative solution framework. Besides, powerful synergies have been revealed between some of the partners, which will be consolidated towards ad-hoc joint commercial exploitation.

7.7 Conclusions

In this chapter, we have presented the CIPSEC project, whose objective is to create a unified security framework that orchestrates heterogeneous, diverse, security products in Critical infrastructure environments. This framework is able to collect and process security-related data (logs, reports, events) so as to generate security anomaly alerts that can affect a CI health and that can have a cascading effect on other CI systems. CIPSEC includes products/tools and services encompassing features such as network intrusion detection, traffic analysis and inspection, jamming attacks detection, antimalware, honeypots, forensics analysis, integrity management, identity access control, data anonymization, security monitoring and vulnerability analysis. The innovation and benefit of CIPSEC relies not only to the addition of all these services and products, but mainly to the integration process of those heterogeneous components has introduced an added value, not covered by the individual solutions, for example allowing to collect all sensors' data of all the products in the XL-SIEM to be analysed, or also allowing to add easily new sensors coming from new future solutions. In summary, the CIPSEC framework integrates all the cybersecurity elements and centralizes all the management in one point, making Critical Infrastructure protection easier to maintain, update and upgrade.

References

[1] Jesse M. Ehrenfeld, 'WannaCry, Cybersecurity and Health Information Technology: A Time to Act.' J Med Syst (2017) 41: 104.

[2] https://ics-cert.us-cert.gov/

[3] CIPSEC project, deliverable D2.2, "D2.2 CIPSEC Unified Architecture First Internal Release", November 2017, https://www.cipsec.eu/content/d22-cipsec-unified-architecture-first-internal-release

[4] CIPSEC project, deliverable D2.5, "D2.5 Final Version of the CIPSEC Unified Architecture and Initial Version of the CIPSEC Framework Prototype", April 2018, https://www.cipsec.eu/content/d25-final-version-cipsec-unified-architecture-and-initial-version-cipsec-framework-prototype

[5] CIPSEC project, deliverable D1.2, "D1.2 Report on Functionality Building Blocks", October 2016, https://www.cipsec.eu/content/d12-report-functionality-building-blocks

[6] CIPSEC project, deliverable D1.1, "D1.1 CI base security characteristics and market analysis", November 2016, https://www.cipsec.eu/content/d11-ci-base-security-characteristics-and-market-analysis

[7] CIPSEC project, deliverable D1.3, "D1.3 Report on taxonomy of the CI environments", November 2016, https://www.cipsec.eu/content/d13-report-taxonomy-ci-environments

[8] CIPSEC project, deliverable D2. 7, "D2.7: CIPSEC Framework Final version".

[9] http://www.hospitalclinic.org/en

[10] https://www.bahn.de/p_en/view/index.shtml

[11] http://www.csipiemonte.it/web/en/

[12] IEEE-SA Standards Board, "IEEE Standard for Software Test Documentation", IEEE Std 829-1998, 16 September 1998.

[13] CIPSEC Deliverable D4.3: "Prototype Demonstration: Field trial results".

8

A Cybersecurity Situational Awareness and Information-sharing Solution for Local Public Administrations Based on Advanced Big Data Analysis: The CS-AWARE Project

Thomas Schaberreiter[1], Juha Röning[2], Gerald Quirchmayr[1], Veronika Kupfersberger[1], Chris Wills[3], Matteo Bregonzio[4], Adamantios Koumpis[5], Juliano Efson Sales[5], Laurentiu Vasiliu[6], Kim Gammelgaard[7], Alexandros Papanikolaou[8], Konstantinos Rantos[9] and Arnolt Spyros[8]

[1]University of Vienna – Faculty of Computer Science, Austria
[2]University of Oulu – Faculty of Information Technology and Electrical Engineering, Finland
[3]CARIS Research Ltd., United Kingdom
[4]3rd Place, Italy
[5]University of Passau, Germany
[6]Peracton, Ireland
[7]RheaSoft, Denmark
[8]InnoSec, Greece
[9]Eastern Macedonia and Thrace Institute of Technology,
Department of Computer and Informatics Engineering, Greece
E-mail: thomas.schaberreiter@univie.ac.at; juha.röning@oulu.fi;
gerald.quirchmayr@univie.ac.at; veronika.kupfersberger@univie.ac.at;
ccwills@carisresearch.co.uk; matteo.bregonzio@3rdplace.com;
adamantios.koumpis@uni-passau.de; juliano-sales@uni-passau.de;
laurentiu.vasiliu@peracton.com; kim@rheasoft.dk;
a.papanikolaou@innosec.gr; krantos@teiemt.gr; a.spyros@innosec.gr

In this chapter, the EU-H2020 project CS-AWARE (running from 2017 to 2020) is presented. CS-AWARE proposes a cybersecurity awareness solution for local public administrations (LPAs) in line with the currently developing European legislatory cybersecurity framework. CS-AWARE aims to increase the automation of cybersecurity awareness approaches, by collecting cybersecurity relevant information from sources both inside and outside of monitored LPA systems, performing advanced big data analysis to set this information in context for detecting and classifying threats and to detect relevant mitigation or prevention strategies. CS-AWARE aims to advance the function of a classical decision support system by enabling supervised system self-healing in cases where clear mitigation or prevention strategies for a specific threat could be detected. One of the key aspects of the European cybersecurity strategy is a cooperative and collaborative approach towards cybersecurity. CS-AWARE is built around this concept and relies on cybersecurity information being shared by relevant authorities in order to enhance awareness capabilities. At the same time, CS-AWARE enables system operators to share incidents with relevant authorities to help protect the larger community from similar incidents. So far, CS-AWARE has shown promising results, and work continues with integrating the various components needed for the CS-AWARE solution. An extensive trial period towards the end of the project will help to assess the validity of the approach in day-to-day LPA operations.

8.1 Introduction

As is the case in other sectors, the problem of securing ICT infrastructures is increasingly causing major worries in local public administration. While local public administrations are, compared to other areas, rarely the target of an attack, using its ICT infrastructure as a springboard for the infiltration of other government systems is of great concern for system administrators. Another significant issue is the danger of becoming a victim of collateral damage ensuing from widespread attacks, as happened to hospitals in the 2017 ransomware attacks [1], causing severe damages to local public administration as well, and going far beyond the loss of reputation. Depending on the criticality of services provided by a local public administration, the damage caused by a successful DDoS, ransomware, malware, or, in the worst case, a destruction-orientated APT attack, can be substantial.

Against this background, the H2020-funded CS-AWARE project[1] aims to equip local public administrations with a toolset allowing them to gain a better picture of vulnerabilities and threats or infiltrations of their ICT systems. This will be achieved via an underlying information flow model including components for information collection, analysis and visualisation which contribute to an integrated awareness picture that gives an overview of the current status in the monitored infrastructure and raises the awareness for both looming and already materialized threats.

Starting from a requirements and situation analysis based on workshops following the soft systems methodology (SSM), Rich Pictures serve as tools for developing a core information flow model that facilitates the information collection, analysis and rendering/visualization processes. In addition to these steps, recommendations are suggested that can either be used as support for human decision makers or are directly executed by (re)configuration scripts to realign defensive capabilities in such a way that existing attacks can be dealt with and developing ones can be prevented from getting through.

In CS-AWARE we develop the building blocks for a cybersecurity awareness solution that builds upon a holistic socio-technological system and dependency analysis. An overview of the proposed approach can be seen in Figure 8.1. After data collection, which is composed of static information collected during system and dependency analysis as well as dynamic information collected at run-time, an analysis and decision support component as well multi-lingual support, will process the information to support the main objectives of the solution:

- Provide situational awareness to system operators or administrators via visualization
- Provide supervised self-healing in cases where the analysis engine could determine an automated solution to prevent or mitigate a detected cybersecurity incident
- Provide the capabilities to share cybersecurity related information with relevant communities to help prevent or mitigate similar incidents for other organizations

To ensure the practical feasibility of the approach, processes and tools developed in this project from the requirements analysis onwards, two city administrations, one medium sized and one large and complex which included outsourced operations, are involved to provide the necessary

---

[1]https://cs-aware.eu/

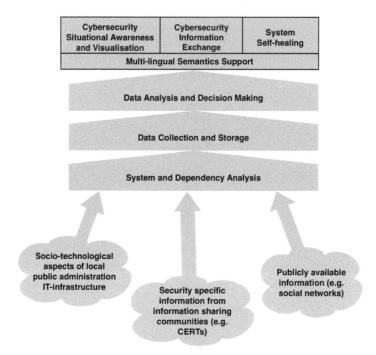

Figure 8.1   The CS-AWARE approach.

guidance and support. Assuming that the pilot implementations are satisfactory at the end of the project, the commercialisation group of the project will then advance the toolset and the services around it into a commercial operation. With the Network and Information Security (NIS) Directive [2] and General Data Protection Regulation (GDPR) [3] having become binding legislation in the European Union in May 2018, it is expected that the need for such a toolset will increase, way beyond local public administration.

The remainder of the chapter is organized as follows: Section 2 discusses related work. Section 8.3 details the CS-AWARE concept and framework, while Section 4 specifies implementation aspects of the main framework components. Section 5 discusses the project results and experiences so far, and Section 6 concludes the chapter.

8.2 Related Work

Cybersecurity affects both individuals and organisations, being one of today's most challenging societal security problems. Next to strategic/critical infrastructures, large commercial enterprises, SMEs and also governmental or non-governmental organisations (NGOs) are affected. Expanding beyond the technology-focused boundaries of classical information technology (IT) security, cybersecurity is strongly interlinked with organisational and behavioural aspects of IT operations, and the need to adhere to the existing and upcoming legal and regulatory framework for cybersecurity. This is particularly true in the European Union, where substantial efforts have been made to introduce a comprehensive and coherent legal framework for cybersecurity. Consequent upon the EU cybersecurity strategy [4], the two main legislatory efforts have been the NIS directive [2] and the GDPR [3]. One of the main aspects of the NIS directive, as well as the European cyber-security strategies is cooperation and collaboration among relevant actors in cybersecurity, as is pictured Figure 8.2 taken from the EU cybersecurity strategy, identifying the main actors relevant for a cooperative and collaborative cybersecurity environment. Enabling technologies for coordination and cooperation efforts are essential for situational awareness and information sharing among relevant communities and authorities. In the long term, it is expected that information sharing can improve cybersecurity sustainably and benefit society and economy in its entirety as an outcome of the enhanced awareness so generated. Current reports such as the 2018 Europol IOCTA (Internet Organised Crime Threat Assessment) [1], support and encourage the growing importance of collaboration and coordination in order to address current and future cybersecurity challenges.

In common with the challenges faced by the NIS, GDPR compliance efforts require greater understanding of an organizations systems in order

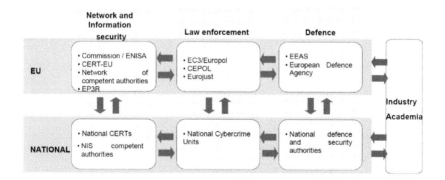

Figure 8.2 Roles and responsibilities in European cybersecurity strategy.

to identify and understand GDPR relevant information and information flows. Awareness technologies like the one proposed in CS-AWARE enable organizations to assess and manage GDPR compliance.

Situational awareness in the CS-AWARE context is a runtime mechanism to gather cybersecurity relevant data from an IT infrastructure and visualise the current situation for a user or operator. Understanding the entirety of the cybersecurity relevant aspects of the internal system is one of the cornerstones for ensuring useful as well as successful collaboration and cooperation between institutions. This is a complex task that will greatly improve the cybersecurity of organisations in the context of cybersecurity situational awareness and cooperative/collaborative strategies towards cybersecurity. Therefore, a system and dependency analysis methodology has been introduced to analyse the environment and

1. Identify assets and dependencies within the system and how to monitor them
2. Capture the socio-technical relations within the organisation and the purely technical aspects
3. Identify external information sources, either official or from dedicated communities
4. Provide the results in an output that can be utilised by support tools

Our work is based on established and well proven methods related to systems thinking, the soft systems methodology (SSM) [5, 6] as well as PROTOS-MATINE [7, 8] and GraphingWiki [9] for system analysis and management/visualization of results. Since technology is only one of many factors in cybersecurity, the system and dependency analysis is designed to detect and analyse the socio-technical nature of an IT infrastructure. It does so by considering the human, organisational and technological factors, as well as other legal/regulatory and business related factors that may contribute to the cybersecurity in a specific context. The key concepts are holism (looking at the entirety of the domain and not at isolated components) and systemic (treating things as systems, using systems ideas and adopting a systems perspective). As can be seen in Figure 8.3, systems thinking is a way of looking at some part of the world, by choosing to regard it as a system, using a framework of perspectives to understand its complexity and undertake some process of change.

Hard and soft systems thinking are the two concepts of systems thinking. Hard systems design is based on systems analysis and systems engineering and it builds on the idea that the world is comprised of systems that can be

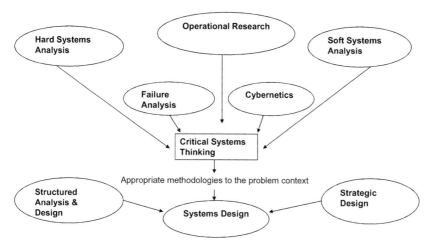

Figure 8.3 Systems thinking.

described and that these systems can be understood through rational analysis. Hard systems design assumes that there is a clear consensus as to the nature of the problem that is to be solved. It is unable to depict, understand, or make provisions for "soft" variables such as people, culture, politics or aesthetics. While hard systems design is highly appropriate for domains involving engineering systems structures that require little input from people, the complex systems and interactions in critical infrastructures or other organisations – especially with cybersecurity in mind – usually do not allow this type of analysis. Soft systems design is therefore much more appropriate and suitable for analysing human activity systems that require constant interaction with, and intervention from people.

Complex systems in software engineering are systems where single components function autonomously but are dependent on the outputs of other components [10, 11] and require abstraction in software engineering can occur in two ways, according to Sokolowski et al. [12] either by limiting the information covered by the model to only the components which are relevant and ignoring the remaining or by reproducing a minimized version of the real-world concept. This procedure of abstraction is critical and sometimes considered one of the most important capabilities of a software engineer [13].

The CS-AWARE modelling approach of the information flow of the complex system is influenced by the Data Flow Diagram (DFD) language defined by Li and Chen [14] but adapted to suit the domains needs. Information flows can cover multiple granularities of interconnections between components, but on a high-level can be classified in three categories: direct information flow, indirect information flow and general information flow [15]. The types of data flows of the original DFD have been adapted, while types of activities were added to ensure that the diversity of the system can be modelled easily. This approach was chosen due to the strong focus and importance of the information flow in the CS-AWARE solution as well as the need for individualised entities.

The role of PROTOS-MATINE and GraphingWiki in this proposed analysis method is to complement the SSM analysis with information from other sources, and provide a solid base for discussion through visualisation in dedicated workshops with the system users and operators. One of the capabilities of GraphingWiki is to instantly link gathered information to other relevant information and thus allow an update of the graphical representation of the analysed system as soon as new information arrives. This feature is used together with SSM analysis to create more dynamic discussions and give even more incentive to the participants to create a system model that is as close to reality as possible.

8.3 The CS-AWARE Concept and Framework

The CS-AWARE framework is the core of the CS-AWARE solution and is based largely on the analysis of cybersecurity requirements for local public administrations and the existing technologies. The aim of the framework is to provide a unified understanding of which components interact with each other and in what way this interaction is made possible. The framework provides a high-level overview of the main components, most of which are represented by one of the consortium partners, as well as a more detailed view of the main subcomponents or processes each of them consist of. Additionally, the relations between these components are defined as well as, in the case of data flows, the data format in which the exchange takes place. The high-level nature of the framework was crucial, since some technical details will only be specifiable during the projects implementation phase.

The CS-AWARE framework consists of an information flow model as well as individual interface definitions for each of the components. The model is a high-level, abstract view on how each of the separate technology components cooperates with the others and in what relation they stand to each other. This might be data flows or also logical control flows between the modules. The focus of the current design of CS-AWARE lies on layers

3, 4 and 7, namely the network, transport as well as application layers of the LPAs systems. To facilitate further analysis, the detailed investigation into the appropriate connections was based on the ETL structured diagram. ETL stands for Extraction, Transformation and Load and is a process most commonly used for database warehouses. Extract stands for the gathering of the data from various sources, Transform for cleaning and manipulating the data to ensure integrity and completeness, Load for transferring the data into its target space [16]. Since the CS-AWARE solution is evidently not a database warehouse, the final layer load was adjusted to better suit the framework's nature and renamed the data-provisioning layer. In our case, the division into layers will mainly be applied to facilitate the structuring of the following, more detailed diagrams of the subcomponents, processes and their interrelationships.

The data extraction layer covers all components responsible for defining relevant data and extracting it, as well as the sources themselves. The system dependency analysis is where the analyst defines relevant sources and data necessary for monitoring the LPAs systems. This information is fed into the data collection module via a control flow, which then extracts the data accordingly.

The data transformation layer summates all components tasked with transforming and analysing the data in some way. The first step is to filter and adapt the data as required before it can be, if necessary, run through the natural language processing information extraction component. The data analysis and pattern recognition and the multi-language support module further process the data. For visualising and sharing the detected incidents and data patterns, the data provisioning layer was defined. This is where all collected information is either visually presented to the end user, shared with selected information sharing communities or used for self-healing rule definition.

The approach chosen to present the CS-AWARE framework interface specifications is based on the classical I/P/O - Input, Process, Output – model, where each component consists of as many input, process and output entities as is required, as visualized in Figure 8.4. For each component, all other building blocks providing data or control flows are summarized as inputs, including which data format they use. Additionally, each component has one or multiple processes or sub components that execute the respective logic of the module and are described in detail as well. Each sub process has inputting and outputting components. Finally, the output components are

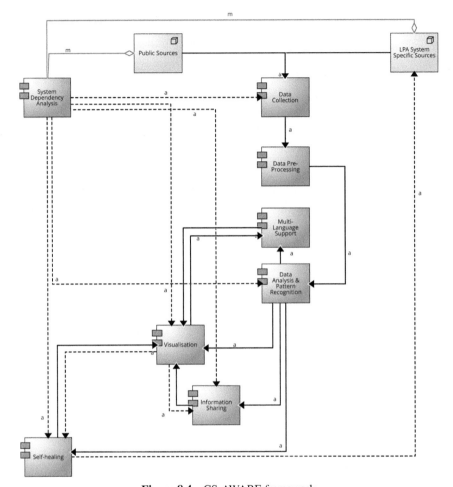

Figure 8.4   CS-AWARE framework.

defined by the same information as the inputs; data format and which type of information flow they use. In preparation of conceptualising the framework, various models and approaches were researched. In the end the CS-AWARE framework was based on the information flows between the components. Nevertheless, it is in line with the NIST cybersecurity framework [17], which identifies five functions as its core: Identify, Protect, Detect, Respond and Recover, making it also compliant with the Italian cybersecurity report, which is based on the NIST framework [18].

Input	Name of Component
Source Module	
	Name of component
Data Format	
	Data format used by component
Description	must cover the following information:
	which type of incoming flow

Output	Name of Component
Destination Module	
	Name of component
Data Format	
	Data format used by component
Description	must cover the following information:
	which type of outgoing flow

Process	Name of Process
Module/s A	
	Name of input components
Module/s B	
	Name of output components
Process	
	Name of process
Definition	
	Description of what happens in this
	subcomponent
Data Input Format	
	Data format used by input component/s
Data Output Format	
	Data format used by output component/s

Figure 8.5 I/P/O interface definition framework.

It was decided that the communication between components illustrated in Figure 8.5, as well as the communication with relevant authorities via the information sharing component will be in accordance with the STIX2 protocol [19]. STIX2 is a modern and flexible protocol to express and link cybersecurity information and is expected to gain wide adoption over the coming years. An open-source java implementation of the protocol specification was developed by CS-AWARE[2] to facilitate wider adoption of the protocol.

---

[2]https://github.com/cs-aware/stix2

8.4 Framework Implementation

This Section discusses in more detail the main framework components identified in Section 3. Section 4.1 discusses the system and dependency analysis approach, Section 4.2 details the data collection and pre-processing steps, Section 4.4 and Section 4.3 discuss the multi-language support and data analysis. In Section 4.5 the visualization component is detailed while Sections 4.6 and 4.7 discuss the information sharing and self-healing components respectively.

8.4.1 System and Dependency Analysis

For analysing the networks and systems in the two European CS-AWARE piloting cities in different countries, one with a population in excess of 2.5 million and a one with a population in excess of 150,000, the Systems Methodology (SSM) was used in conjunction with GraphingWiki. The two cities participated in the project as pilot use cases for whom cybersecurity awareness systems were to be built as an output of the project.

The two cities presented very different problem domains: one city's system was extremely large, reflecting as it did the size of the population it served and potentially had 15+ million concurrent citizen users. These users can access the city's systems both from their homes, public buildings and wireless hotspots around the city. This city has outsourced the management many of its key systems. The network topology, the systems and underlying process combine to form what overall is an extremely complex system. The size and complexity of the system precludes any one individual, or indeed small group of employees' form having a complete understanding of all of the systems or the links between systems and their processes and sub-processes. The smaller city operates, manages and maintains all of its own systems.

SSM is a well-proven analytical approach to systems analysis that has been used in an extremely wide range of settings. It is beyond the scope of this chapter to give anything but a brief description of the methodology.

SSM consists of seven stages:

1. Enter the problem situation
2. Express the problem situation
3. Formulate root definitions of systems behaviour
4. Build conceptual models of systems in root definitions
5. Compare models with real-world situations
6. Define possible and feasible changes
7. Take action to improve the problem situation

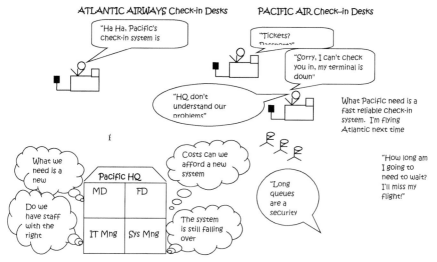

Figure 8.6   Soft systems analysis rich picture.

The problem situation is explored (expressed), by drawing "Rich Pictures". These pictures are cartoon-like representations that are intended to encompass all of the elements of the situation being examined, be they technical, social, economic, political. A machine drawn example can be seen in Figure 8.6, and depicts a malfunctioning airline passenger check-in system and outlines different viewpoints of those involved when one airline's check-in systems fails.

The analysis of both of the city's operation was conducted in the first two of a proposed series of three workshops. In the first workshops, the participants were asked to draw rich pictures to identify their city's key critical systems (those systems critical to their city's ability to provide services to its citizens and those systems storing or processing sensitive or personal information). Having identified the critical systems, further rich pictures were drawn to explore the interrelationships between the systems so identified in terms of network connectivity and information flows.

These rich pictures informed the development of a series of GraphingWiki graphs, like the one seen in Figure 8.7, which enabled the analysts to represent and model their understanding of the networks and systems in both pilot cities. Each of the nodes is a wiki page that holds the semantic descriptions of the respective elements.

A second round of workshops in the pilot cities was undertaken in which the analysts decided to use the CATWOE approach [20] to gain a better

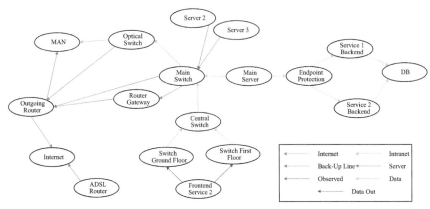

Figure 8.7   System and dependency analysis use case example.

understanding of the processes depicted in the rich pictures created during the first workshops. CATWOE (a mnemonic) was used to identify, express explore and explain the following features in the key rich pictures drawn in the workshops. In doing so, the participants described the processes and sub-processes of the key systems identified in the first and second workshops.

Customers	The organisations customers. The stakeholders of the system
Actors	The employees of the organisation. The people involved in ensuring that a transformation takes place
Transformation	The process by which inputs become outputs e.g. raw materials become finished goods
World view	The wider view of all of the interested parties – employees, suppliers, customers etc... The "big picture"
Owner	The owner of the system or process. The organisation in control
Environmental constraints	Finances, legislation ethics

These CATWOE analyses were then used in a plenary session to further correct and refine the representation of the systems as mapped out in the GraphingWiki, and allowed to identify the information flows through the systems each of those processes produce during day-to-day operations. The identification of information flows is considered a key aspect of understanding where and how to best monitor the systems in the cybersecurity context and are the key to interface the analysed systems with CS-AWARE.

8.4.2 Data Collection and Pre-Processing

For data collection and pre-processing the main challenge in CS-AWARE is to deal with the diversity of data collected from various sources, ranging from cybersecurity information that is heavily structured (e.g. STIX based information sources), to loosely structured information (e.g. log files) or completely free semantic text (e.g. social media). It was decided to convert incoming data from all sources to STIX2 format in the pre-processing stage.

Data collection and pre-processing are applied to multiple sources and the retrieved data is stored in a data-lake. To handle large volumes and a variety of file formats, a big-data pipeline has been implemented following a flexible approach, so that data sources can easily be integrated at a later stage, should additional relevant data sources be identified. Importantly, the collection has been executed in compliance of GDPR regulation where personal data are removed or anonymized at source, since personal data is not required for CS-AWARE operation in the majority of use cases. The implemented framework aggregates and ingests three main classes of information sources:

- Logs from servers, databases, applications and network/security devices from within monitored systems
- Cyber threat intelligence from specialised websites and feeds
- More general cybersecurity related notifications and warnings collected from social networks

In order to collect information from the monitored systems within the local public administrations that usually do not have APIs for data collection, a collector interface was developed to be hosted within the monitored systems. It acts as a local collector of data that is relevant for CS-AWARE and provides an interface to the CS-AWARE solution which may be hosted in the cloud. The conversion to STIX2 format is usually straight forward, because the relevant information is often based on unusual behaviour which can be easily modelled in STIX2.

Threat intelligence sources usually provide a public API that allows collection of data, but there is no agreed or standardized data format in which this data is provided. Common formats are among others STIX1/STIX2, comma separated values (CSV), eXtensible Markup Language (XML) or Java Object Notation (JSON). Since CS-AWARE operates on STIX2, all collected data entries are converted to STIX2 in data pre-processing. Since almost exclusively information with a strong cybersecurity context is shared by threat intelligence sources, the conversion is usually straight-forward.

Threat intelligence notifications collection is performed every 12 hours and stored within the CS-AWARE repository.

As part of this project we want to explore the opportunity of cybersecurity prevention and notifications by listening to social media sources such as Reddit and Twitter. The intuition here is that a cyber-attack may propagate following a certain pattern that could be anticipated by social media warnings, and social media conversations often provide an early indicator to information that may be shared by threat intelligence at a later time. A challenge with utilizing unstructured semantic text like social media is to assess the relevance of each element and assign a structure to it so that it can be processed in an automated way. In CS-AWARE we try to answer this challenge with a natural language processing (NLP) based information extraction approach that will be discussed in more detail in Section 4.3.

As project repository we believe that a winning approach would be a cloud based big-data repository since it offers a ready-to-use framework designed to scale up in a cost effective manner. For this type of challenge, a popular approach involves using a queue system, such as Apache Kafka, and a database where the data is stored; this infrastructure could be well replicated on major cloud providers. Having said that, a full functioning big-data pipeline has a fixed cost even if not fully exploited. For this reason, we preferred a slim and flexible solution where costs are compressed. In more detail, we created the CS-AWARE data-lake on AWS S3[3] storage. AWS S3 provides capabilities to store and retrieve any amount of data from anywhere. It is worth mentioning that thanks to a structured folder hierarchy, it is intuitive and straight forward to retrieve the needed information. Despite the low cost and simplicity, this approach already demonstrated to be fast and stable.

8.4.3 Multi-language Support

In CS-AWARE, the existing technology to support handling of multiple languages is used and has been adopted to fit specific needs of the project context and the use cases. To this aim *Graphene*, a rule-based information extraction system developed in the context of research conducted at the University of Passau, was utilized. There are two use-cases for multi-language support in CS-AWARE: multi-language support at the input when cybersecurity relevant information is collected from multiple sources, and

---

[3]https://aws.amazon.com/s3/

multi-language support at the output to inform the system operators of the systems security status in their chosen language. In this Section we focus on the first use case, where the challenge is not only to translate new incoming information to a meta-language, but at the same time to extract the most relevant information using natural language processing (NLP) methods.

In the project framework, Graphene is responsible for all functions of the NLP information extraction component. The tool uses a two-layered transformation stage consisting of a *clausal disembedding layer* and a *phrasal disembedding layer*, together with *rhetorical relation identification*. To put this in simpler terms, the main approach we take here is to *simplify complex sentences* before applying a set of tailored rules to transform a text into the knowledge graph. During the CS-AWARE project, we had the opportunity to mature the original research prototype as a technology which is both easy to deploy as a service and integrate as a product using de-facto web standards. Additionally, we also had the opportunity to implement and add a new extraction layer responsible for transforming complex categories – what one would call 'coarse-grained information' – into a graph of fine-grained knowledge, as described in the implementation section.

Consequent upon Graphene's ability to extract complex categories, we are able to extract useful information in the correct level of granularity. As an example, we consider the case of a recent tweet written by the United States Computer Emergency Readiness Team (US-CERT), as shown in Figure 8.8.

Once we remove the links and hashtags, the knowledge graph generated from Graphene allows us to identify vendor and products that might be under attack or suffering from new vulnerabilities. With this functionality, both types of information can be forwarded to users and system admins as quickly as they are published in a social network like Twitter. More elaborate information and technical details about the information extraction strategy, including the sentence simplification step and the identification of the rhetorical structures can be found in [21], while for the extraction of complex categories more elaborate information can be found in [22].

8.4.4 Data Analysis

One of the main tasks of the CS-AWARE platform is to look for various threat patterns some of which may have not been detected or recognised as such before and which can signal either a clear threat or a suspicious behaviour that may possibly or potentially be a threat. The way we define a threat pattern at a conceptual level that it is considered as an open set of individual threat parameters with unique settings/values aimed to capture anomalous events.

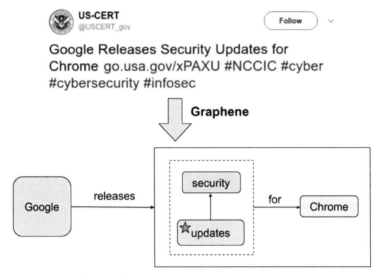

Figure 8.8 From tweets to knowledge graphs.

Such a set of threat parameters can be altered and improved with time as the knowledge about threats expands. Once such patterns catch an occurrence of multiple suspicious events simultaneously, then the identified events are flagged for further analysis. Many times one suspicious event may not be enough to be considered a threat but when multiple suspicious events happen, then the chances to have a threat increases.

In light of the above, we take as data analysis as being the set of processes where all data sources are assembled, combined and searched for unusual or threat patterns. Handling the data sources, their format and the way they should be cleaned from overhead, prepared and then analysed is vital for finding unusual threats or patterns that otherwise may go undetected by the existing tools in the market. Our data analysis efforts are focused on internal data sources belonging to organizations that use the CS-AWARE platform, such as logs, as well as external data sources such as threat intelligence platforms, specialized cyber-security forums, news and solutions. Such data is in a raw form and will have to be filtered and processed in order to extract only the most useful data for analysis.

The data analysis focuses on extracting the most probable elements of information that could form cyber security threats as well as info related to such threats. The data analysis that builds on a Peracton MAARS^{TM} component combines the above sources to identify threats and possible security incidents. Some combinations, assuming a proper information pre-processing, will be quite straightforward to process while others might need some more advanced analysis.

In this respect, the data analysis engine should be able to perform at least the following with regards to the above sources:

- *Match vulnerability information to assets* – e.g. a vulnerability found on a specific OS version; is it applicable to monitored LPAs.
- *Combine threat information with logs and assets* – the analysis should be done based on specific attributes that characterize an attack e.g. to identify a security incident regarding suspicious activity originating from a specific IP and targeting specific systems there is a need to match these attack characteristics to the information we have, i.e. we have to analyse threat information provided by external sources to give values to these attributes, and once we do so, process LPA's logs and LPA's assets inventory to identify these.
- *Attack pattern matching* – analyse network and system activity to identify potential security incidents based on attack patterns either collected from external sources or specified by CS-AWARE security experts. The engine's efficiency strongly depends on the defined patterns. Although the engine should demonstrate its ability through a pre-defined set of patterns, it should also be able to accommodate additional patterns that security experts would like to define in the future.

We expect that the data analysis should provide information about the criticality of the specific security incident and, based on this classification, suggest the most appropriate risk mitigation option (if available from data). Revisiting the above example where a threat that reports suspicious activity originates from specific IPs and targets specific systems, the (risk) analysis for the following scenarios will give us the corresponding results described below:

- *Scenario 1*: Threat is flagged by external sources as critical and the LPA has systems that are vulnerable to this malicious activity: the risk for the organization is high. A risk mitigation strategy should be applied, i.e. an action is required to mitigate the risk, the details of which are subject to the information provided by external sources or by a CS-AWARE security expert.
- *Scenario 2*: Threat is flagged by external sources as critical, logs indicate incoming traffic from this IP, yet the LPA has *no* systems that are vulnerable to this malicious activity: the risk is low. In this case, *no* action is required.

8.4.5 Visualization

The visualization component will show the users (e.g. system administrator, management) the level of cyber threats to their system and will make it possible for system administrators to cooperate with the system to identify self-healing procedures and to share information with external partners regarding new cyber threats that have been identified in the analytics module.

The visualisation module is also the main user interface of the CS-AWARE product for administration. In order to provide cybersecurity awareness, it is necessary to visualise the threats, the threat level, the possible self-healing strategies and the information shared with the cybersecurity community. It is also necessary to have an interface to communicate back to the system information regarding controlling the aforementioned topics as well as lower level administration. The visualisation component will take care of this according to the work done in the dependency analysis and in good cooperation with other parts of the CS-AWARE solution. For the interchange of data, the STIX2 format has been determined as being the basic communication format between the modules, as it is commonly used in the field of cybersecurity and is both fairly stable, extensible when needed and with a reasonable support of frameworks with which to work.

The number of cybersecurity events has also been rising over the past years and as more and more of our society is based on information systems, the issues have multiplied over time. Before this project, a number of independent vendors have different visualisation means to show how their particular system is threatened by cyber security events. In a large-scale facility like most LPAs this results in a large number of reports on what is going on in their field of operation. For the system administrator this only gives a partial overview of the cyber security events, as it on one hand only delivers the view from the single vendor, and on the other hand often is too complex to be useful. The number of different reports to choose from can be high and they are usually only collected per vendor. This makes it difficult to assess the full cyber security overview. The paucity of overview leaves the cybersecurity awareness level lower than it could be. This is a situation that needs to be remedied.

The main gap is that current systems lack a significant cognitive component, in order to propagate the overall level of cybersecurity awareness. Specifically, we have identified the lack of single point of overview, and together with the rising level of entropy in cybersecurity reporting, which is believed to be the consequence of the multitude of sources that may not be connected. This both results with information overload and the already

Figure 8.9 The CS-AWARE visualization component.

mentioned lack of common cybersecurity awareness. The main solution is to have a single interface to propagate the immediate cybersecurity awareness situation to the system administrator and other users who have a need for this information. For this we have developed a dashboard that – in an early version – is shown in Figure 8.9. It makes it easy to overview all concurrent cybersecurity threats and vulnerabilities as well as a summarised threat level. Through the dashboard, the system administrator has direct access to self-healing strategies, suggestions as well as possible information sharing texts on newly found threats.

We are generating a single view of cybersecurity threats and vulnerabilities that will show all of the major threat types and the summarised threat level. These will be shown over time to help understand the urgency and how the change in threat level is evolving, in order to mitigate a threat in the best way at that time. A reduction in time spent looking for a cybersecurity issue is worth many hours of post-mortem issue fixing and cleaning. Notice that the dashboard will have a graphic that continuously shows development over time in both size and colour, in order to let the system administrator act swiftly and becoming aware of cybersecurity events much faster than going through heaps of internet pages to find a possible change.

The visualisation component has interfaces to the system and dependency analysis, the data analysis and pattern recognition as well as the multi-language (NLP) support components, and also to the self-healing and information sharing components, where information sharing to the cybersecurity communities will be for the common good. This way the visualisation component enhances the cybersecurity awareness and helps the system administrators maintain their systems unaffected through a faster and better decision making and self-healing process.

8.4.6 Cybersecurity Information Exchange

The CS-AWARE cybersecurity information exchange (CIE) provides a dissemination point for cyber threat information (CTI) that CS-AWARE components have collected, analysed, identified and classified as "shareable". It is the interface to external entities, such as Computer Emergency Response Teams (CERTs), Computer Security Incident Response Teams (CSIRTs) and other threat intelligence platforms to inform them about threats, sightings (i.e. an observation related to a threat) and mitigation actions. This information will be mainly produced by the CS-AWARE data analysis component or by external sources and enhanced by the former.

CTI is information that is constantly generated and shared among devices and departments within (especially large) organisations which have well-established procedures for appropriately handling sensitive, classified and personal information found within CTI. When CTI is about to be shared with external entities, several interoperability and security issues have to be confronted [23], which can be categorised according to the three layers depicted in Figure 8.10.

Although the *legal* framework might encourage or require the sharing of cyber-threat information, as the NIS directive does, several other legal requirements might prohibit or restrict the uncontrolled sharing of CTI. One of the main legal restrictions arises from the GDPR and relates to the personal information shared with external entities without the user's consent. In the case of CTI, personal data might be part of the shared sightings, such as IP addresses or usernames of entities that have been identified as sources of malicious activity. CTI sharing with external entities should not impact privacy and personally identifiable information (PII), and therefore, data

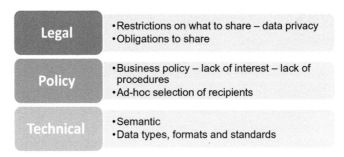

Figure 8.10   CTI exchange interoperability layers.

anonymization should take place if necessary, prior to sharing CTI with external entities or being made public. However, certain data that under certain circumstances might be considered as personal (e.g. IPs), are very important for the receiving parties to have. Otherwise the information provided becomes useless, and therefore should be excluded from any anonymization processing. Moreover, based on Article 49 of the GDPR, the processing of personal data by certain entities, such as CERTs and CSIRTs, strictly for the purposes of ensuring network security is permitted as it constitutes a legitimate interest of the data controller.

An organization's *policy* should address issues related to information sharing, while well-established procedures and appropriately deployed measures will help avoid the leakage of classified or sensitive information. Data sanitisation [24] is one of the solutions that the organisations should consider utilising to ensure that no sensitive or classified information is disclosed to unauthorised entities while sharing CTI with external entities. Policy restrictions with regards to sharing should also be supported by appropriate technical measures.

On the *technical* layer, adoption of standardised schemes used for sharing cyber-threat information is deemed necessary to achieve the necessary semantic and technical interoperability. The STIX2 protocol is the information model and serialisation solution adopted by CS-AWARE for the communication and sharing of CTI.

STIX2 also supports data markings which can facilitate enforcement of policies regarding the sharing of information. More specifically, STIX2 supports statements (copyright, terms of use, . . .) applied to the shared content as well as the Traffic Light Protocol (TLP)[4,5] (a set of designations used to ensure that sensitive information is shared with the appropriate audience by providing four options as shown in Figure 8.11). Although optimized for human readability and person-to-person sharing and not for automated sharing exchanges, the adoption of TLP in CS-AWARE will help restrict information sharing only with specific entities or platforms and avoid any further unnecessary or unauthorized dissemination thereof.

Considering the limitations of TLP which cannot support fine-grained policies, the CS-AWARE information exchange component also adopted the Information Exchange Policy (IEP), a JSON based framework developed by

---

[4]https://www.first.org/global/sigs/tlp/
[5]https://www.enisa.europa.eu/topics/csirts-in-europe/glossary/considerations-on-the-traffic-light-protocol

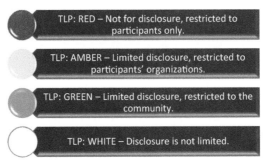

Figure 8.11 The traffic light protocol.

FIRST IEP SIG (2016) [25]. IEP is not supported in the current version of STIX, yet STIX compatibility was considered in its design.

8.4.7 System Self-Healing

Self-healing is described as the ability of systems to autonomously diagnose and recover from faults with transparency and within certain criteria. Although these criteria vary according to the system's infrastructure, they often include requirements such as availability, reliability and stability [26]. Self-healing constitutes an important building block of the CS-AWARE architecture, which aims to assist LPA administrators in responding to identified vulnerabilities and high-risk threats by providing customised healing solutions or recommendations. The self-healing component is an innovative fully-supervised solution that uses the results of the analysis performed by the analysis component. The latter processes cyber threat information collected from external sources, internal logs and LPA architecture specifics and produces knowledge about potential high-risk situations for a specific LPA. Based on the aforementioned outcome, self-healing looks for the most appropriate mitigation solution among those provided by the external sources or found in the self-healing enhanced database of appropriate solutions. Supervision is defined as the degree of required human interaction concerning the feedback mechanism and the expansion of self-healing mechanisms [26]. Self-healing systems are categorised as fully supervised, semi-supervised or unsupervised. Figure 8.12 provides an overview of the research work accomplished on self-healing properties as published in [27].

The composed mitigation rules aim to enhance the availability and overall security of the system while simultaneously reducing the required

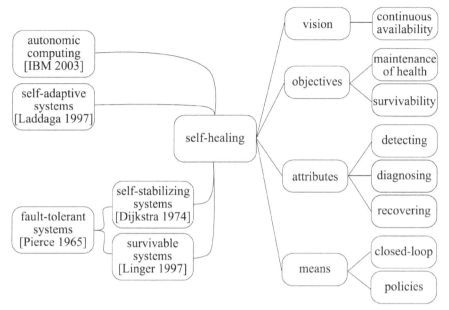

Figure 8.12 Properties of self-healing research.

workload of system maintainability. Furthermore, the CS-AWARE self-healing component has the ability autonomously to diagnose and mitigate threats, while ensuring that the system's administrator, who is always aware of the system behaviour, can prevent configuration changes that may raise incompatibility issues.

The self-healing component also interacts with the visualisation component for the following purposes:

- inform administrators about mitigation solutions applied to LPA systems
- request LPA administrator permission to apply a solution
- provide recommendations about how to confront an identified high-risk situation or vulnerability

The self-healing component is fully supervised, always allowing the LPA administrator to decide whether or not they want to apply the suggested mitigation rule. It utilises the results of the data analysis component provided in STIX2. Once the self-healing receives input data from the analysis component, it identifies the threat type and composes the proper mitigation rule autonomously. Rules composed by the self-healing module incorporate three alternatives:

- Inform LPAs about which acts to perform in order to avoid the threat or reduce the impact (recommendation)
- Ask for the LPA administrator's permission in order to apply the rule automatically
- Automatically apply the rule, provided that the administrator has set this preference through the visualisation component

The self-healing component consists of three main and three auxiliary subcomponents, whose interaction is shown in Figure 8.13. The main subcomponents were defined in the CS-AWARE framework while auxiliary subcomponents were defined during the design phase to facilitate the composition and application of mitigation rules. The self-healing policies are a database which contains records of potential threats that might be detected in an LPA system and the corresponding mitigation rules. The mitigation rules are stored in a human-readable generic format as well as in machine-readable format. Moreover, the self-healing policies subcomponent includes entries which contain the CLI syntax of LPAs central nodes.

The decision engine initiates the composition of a rule. It performs searches of the self-healing policies database for a matching rule. If it finds a match, it initiates a rule which is in a human-readable format. The security rules composer accepts input from the decision engine subcomponent in a human-readable format and converts it to a machine-readable format based on the CLI syntax of the affected node. The parser parses the STIX package

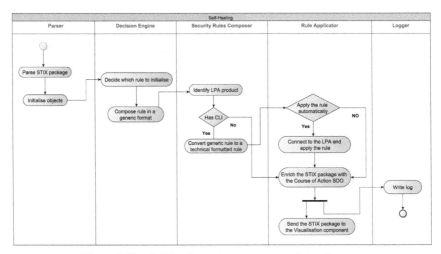

Figure 8.13   Self-healing subcomponents activity diagram.

and extracts useful data for the process of mitigation rules composition, and the rule applicator is responsible for enriching the STIX package with the mitigation rule, sending data to the visualisation component and for applying the rule on the remote machine. In case the mitigation rule must be applied remotely then the self-healing connects to the remote node and applies the rule automatically provided that the LPA administrator has given permission. Finally, the logger writes a log entry in the log file which contains information about how the mitigation rule was applied.

8.5 Discussion

The demand for cybersecurity tools is strong. An alarming rate of purposeful cyber-attacks forces authorities on different levels to do more than just to be reactive operations. At the same time new regulatory and legal requirements are implemented by the highest-level authorities and are effecting how systems can be operated and data can be handled on the regional level. In Europe, especially the NIS directive is concerned with how the most critical services for our society are handling cybersecurity, while the GDPR is protecting an individual person's information and privacy. This has caused actions and worries with private companies but is affecting also many functions of local public administrations. Although the local public administrations have not been the direct targets of malware attacks they are crucial providers of services governing our everyday lives and are heavily influencing society on a regional level. The CS-AWARE project has proven to be even more current and relevant than we could have anticipated during the time of writing the proposal.

The first year of the project has been successful. Two rounds of dependency analysis workshops at our piloting municipalities have been completed and have provided extensive insight into the operations of local public administrations. We have gained valuable information that has influenced and guided the CS-AWARE framework development and implementation. We have seen that there are substantial differences in LPA operations between countries even on the European level. Besides the obvious differences in language in national and regional levels in Europe, we have seen that the rules and regulations guiding LPA operations are substantially different between countries, and may affect how cybersecurity tools like the CS-AWARE toolset can be deployed and operated. We have also seen however that the CS-AWARE concept and framework is well suited to handle these differences due to the flexible and socio-technological analysis at its base. We believe that we have proven the framework to be valid. It is now modified and adjusted based on knowledge and circumstances derived from the LPA use cases. The project will continue with the framework implementation and integration, and an extensive piloting phase towards the end of the project will allow us to draw broader conclusions about the usefulness of cybersecurity awareness technologies in day-to-day operations of local public administrations.

An important lesson we have already learned at this stage is how important collaboration and information sharing are. Cooperation and collaboration is essential and becoming more relevant in future, since there are many actors on the local public administration level. Small cities and communes with individually centralized organizations, but each distributing responsibilities among external experts. The larger the commune, the greater appears to be the silo effect. Then even a single service forms, an isolated unit which does not have direct collaboration with other city services. Information sharing is therefore a key factor to generate and understand the full picture of the internal infrastructures. While our information sharing efforts were focused on sharing cybersecurity information with external authorities, such as the NIS competent authorities listed in Figure 8.2, we have seen that in practice already sharing with different actors on the local level (other departments or suppliers) may have a significant positive effect on cybersecurity. This is one aspect that will be more closely looked into during the piloting phase of CS-AWARE. We are investigating this even further in another H2020 project, CinCan (Continuous Integration for the Collaborative Analysis of Incidents)[6], where we also try to promote sharing and reporting vulnerability information between different countries' CERT organizations.

We feel that CS-AWARE is not just an individual project, but a continuous path we need and have now started to follow. Technology touches every aspect of our lives and we need tools that allow us to safely utilise them by covering all legal security requirements.

8.6 Conclusion

In this Chapter we have presented the EU-H2020 project CS-AWARE (running from 2017 to 2020), aiming to provide cybersecurity awareness technology to local public administrations. CS-AWARE has several unique features, like the socio-technological system and dependency analysis at the

---

[6]https://cincan.io/index.html

core of the technology that allows a fine grained understanding of LPA cyber-security requirements on a per case basis. Furthermore, the strong focus on automated incident detection and classification, as well as our efforts towards system self-healing and cooperation/collaboration with relevant authorities are pushing the current state-of-the-art, and are in line with cybersecurity efforts on a European and global level.

In light of a substantially changing legal cybersecurity framework in Europe, we have shown that CS-AWARE is an enabling technology for many cybersecurity requirements imposed by these regulations. For example, information sharing of cybersecurity incidents is a requirement of the NIS directive for organizations classified as critical infrastructures, and may in future be extended to other sectors as well. Similarly, the identification of personal information and information flows within organizations systems, as done in the system and dependency analysis of CS-AWARE, is a key requirement for GDPR compliance.

We have detailed the CS-AWARE framework and have shown how the different building blocks are implemented in CS-AWARE. We have discussed the first results of the project, especially the outcomes of two rounds of system and dependency analysis workshops in the piloting municipalities of CS-AWARE, and we have discussed how those results are influencing the framework implementation and integration in preparation for the piloting phase of the project. Our initial results show the necessity of awareness technologies in LPAs. Administrators and system operators are looking for solutions that improve awareness of cybersecurity incidents on a system level and assist with prevention or mitigation of such incidents. We have seen a specific need for awareness as well as improved collaboration and cooperation between different departments or suppliers, an area that is often neglected but has significant potential for introducing cybersecurity risks.

CS-AWARE will continue with further developing of the technological base and integration of the components that form the CS-AWARE framework. An extensive piloting phase towards the end of the project will give insights into the practical feasibility and relevance of the awareness generating tech-nologies, and allow us to evaluate how both system administrators and system users can benefit from CS-AWARE. The piloting phase will be accompanied by social sciences based study to evaluate how the CS-AWARE technologies are accepted by its users in day-to-day operations. At the same time, we will continue to promote CS-AWARE among potential users, implementers and authorities to bridge the gap between legal and regulatory requirements and actual technology that can fulfil those requirements. In an era where it is thought that cybersecurity can only be effective through cooperation and collaboration, constant interaction between the main actors is important to achieve a comprehensive and holistic solution.

Acknowledgements

The authors would like to thank the EU H2020 project CS-AWARE (grant number 740723) for supporting the research presented in this chapter.

References

[1] Europol, "Internet Organized Crime Threat Assessment (IOCTA) 2018," Online, 2018.

[2] THE EUROPEAN PARLIAMENT AND THE COUNCIL OF THE EUROPEAN UNION, DIRECTIVE (EU) 2016/1148 OF THE EUROPEAN PARLIAMENT AND OF THE COUNCIL *of 6 July 2016 concerning measures for a high common level of security of network and information systems across the Union,* 2016.

[3] THE COUNCIL OF THE EUROPEAN UNION, REGULATION (EU) 2016/679 OF THE EUROPEAN PARLIAMENT AND OF THE COUNCIL of *27 April 2016 on the protection of natural persons with regard to the processing of personal data and on the free movement of such data, and repealing Directive 95/46/EC,* 2016.

[4] European Commission and High Representative of the European Union for Foreign Affairs and Security Policy, *Cybersecurity Strategy of the European Union: An Open, Safe and Secure Cyberspace,* JOIN(2013) 1 final, 2013.

[5] P. Checkland, "Systems Thinking, Systems Practice," Wiley [rev 1999 ed], 1981.

[6] P. Checkland, "Soft Systems in Action," Wiley [rev 1999 ed], 1990.

[7] P. Pietikainen, K. Karjalainen, J. Röning and J. Eronen, "Socio-technical Security Assessment of a VoIP System," in 2010 *Fourth International Conference on Emerging Security Information, Systems and Technologies,* 2010.

[8] T. Schaberreiter, K. Kittilä, K. Halunen, J. Röning and D. Khadraoui, "Risk Assessment in Critical Infrastructure Security Modelling Based on Dependency Analysis," *in Critical Information Infrastructure Security:*

6th International Workshop, CRITIS 2011, Lucerne, Switzerland, September 8–9, 2011, Revised Selected Papers, 2011.

[9] J. Eronen and J. Röning, "Graphingwiki – a semantic wiki extension for visualising and inferring protocol dependency," in First Workshop on *Semantic Wikis – From Wiki To Semantics, 2006.*

[10] J. Jiang, J. Yu and J. Lei, "Finding influential agent groups in complex multiagent software systems based on citation network analyses," *Advances in Engineering Software,* pp. 57–69, 2015.

[11] L. Saitta and J.-D. Zucker, Abstraction in artificial intelligence and complex systems, Springer, 2013.

[12] J. Sokolowski, C. Turnitsa and S. Diallo, "A conceptual modeling method for critical infrastructure modeling," *in Simulation Symposium,* 2008. *ANSS 2008. 41st Annual,* 2008.

[13] J. Kramer, "Is abstraction the key to computing?," *Communications of the ACM,* pp. 36–42, 2007.

[14] Y.-L. Chen and Q. Li, Modeling and Analysis of Enterprise and Information Systems: from requirements to realization, Springer, 2009.

[15] P. Clemente, J. Rouzaud-Cornabas and C. Toinard, From a generic framework for expressing integrity properties to a dynamic mac enforcement for operating systems, Springer, 2010, pp. 131–161.

[16] S. Bansal and S. Kagemann, "Integrating big data: A semantic extract-transform-load framework," *Computer,* pp. 42–50, 2015.

[17] NIST National Institute of Standards and Technology, "Framework for Improving Critical Infrastructure Cybersecurity," 2015.

[18] CINI Cyber Security National Laboratory, "Italian Cyber Security Report 2015 – A National Cyber Security Framework," 2016.

[19] OASIS Committee Specification 01, STIX Version 2.0. Part 1: STIX Core Concepts, R. Piazza, J. Wunder and B. Jordan, Eds., 2017.

[20] D. S. Smyth and P. B. Checkland, "Using a Systems Approach: The Structure of Root Definitions," *Journal of Applied Systems Analysis,* vol. 6, no. 1, 1976.

[21] M. Cetto, C. Niklaus, A. Freitas and S. Handschuh, "Graphene: Semantically-Linked Propositions in Open Information Extraction," in *In Proceedings of the 27th International Conference on Computational Linguistics (COLING),* New-Mexico, USA, 2018.

[22] J. E. Sales, A. Freitas, B. Davis and S. Handschuh, "A Compositional-Distributional Semantic Model for Searching Complex Entity Categories," in *5th Joint Conference on Lexical and Computational Semantics (*SEM),* Berlin, 2016.

[23] C. S. Johnson, M. L. Badger, D. A. Waltermire, J. Snyder and C. Sko-rupka, "Guide to Cyber Threat Information Sharing," *National Institute of Standards and Technology,* pp. NIST SP 800-150, 2016.

[24] M. Bishop, B. Bhumiratana, R. Crawford and K. Lwvitt, "How to sanitize data?," in *13th IEEE International Workshops on Enabling Technologies: Infrastructure for Collaborative Enterprises,* Modena, 2004.

[25] Forum of Incident Response and Security Teams (FIRST), "Information Exchange Policy Framework, Version 1.0".

[26] C. Schneider, A. Barker and S. Dobson, "A survey of self-healing systems frameworks," *Software: Practice and experience,* vol. 45, no. 10, pp. 1378–1394, 2014.

[27] H. Psaier and S. Dustdar, "A survey on self-healing systems: approaches and systems," *Computing,* vol. 91, no. 1, p. 47, 2010.

9

Complex Project to Develop Real Tools for Identifying and Countering Terrorism: Real-time Early Detection and Alert System for Online Terrorist Content Based on Natural Language Processing, Social Network Analysis, Artificial Intelligence and Complex Event Processing

Monica Florea[1], Cristi Potlog[1], Peter Pollner[2], Daniel Abel[3], Oscar Garcia[4], Shmuel Bar[5], Syed Naqvi[6] and Waqar Asif[7]

[1]SIVECO Romania SA, Romania
[2]MTA-ELTE Statistical and Biological Physics Research Group, Hungary
[3]Maven Seven Solutions Zrt., Hungary
[4]Information Catalyst, Spain
[5]IntuView, Israel
[6]Birmingham City University, United Kingdom
[7]City, University of London, United Kingdom
E-mail: Monica.Florea@siveco.ro; cristi.potlog@siveco.ro;
pollner@angel.elte.hu; daniel.abel@maven7.com;
oscar.garcia@informationcatalyst.com; sbar@intuview.com;
Syed.Naqvi@bcu.ac.uk; Waqar.Asif@city.ac.uk

In the last decades, the importance of social media has increased extremely with the creation of new communication channels and even changing the way people are communicating. These trends came along with the disadvantage of allowing a new scenario where messages containing valuable data about critical threats like terrorism and criminal activity are ignored, due to the sheer inability to process – much less analyze – the vast amount of available data. Terrorism has a very real and direct impact on basic human rights of victims, such as the right to life, liberty and physical integrity, often with devastating consequences.

In this context, the RED-Alert project was designed to build a complete software toolkit to support LEAs in the fight against the use of social media by terrorist organizations for conducting online propaganda, fundraising, recruitment and mobilization of members, planning and coordination of actions, as well as data manipulation and misinformation. The project aims to cover a wide range of social media channels used by terrorist groups to disseminate their content which will be analysed by the RED-Alert solution to support LEAs to take coordinated action in real time but having as a primordial condition preserving the privacy of citizens.

9.1 Introduction

Radicalisation leading to violent extremism and terrorism is not a new phenomenon but the way it is now spreading is more and more alarming and extending to the EU as a whole. As a matter of urgency, the European and Member States' policies must evolve to match the scale of the challenge offering effective responses [1].

During recent years Europe is facing new challenges to design and build new tools and to take advantage of technological advancements to prevent terrorist attacks. The Europol report from 2017 shows that, in 2016, a total of 142 failed, foiled and completed attacks have been reported. In 2017, 16 attacks struck eight different Member States while more than 30 plots were foiled.[1]

The RED-Alert project is aligned to SECURITY Work Programme 2016–2017 call objectives that targets improvement of investigation capabilities, solving crimes more rapidly, reducing societal distress, investigative costs and the impact on victims and their relatives and to prevent more terrorist endeavours.[2]

The RED-Alert project is a H2020 European research and development project that uses analytics techniques such as NLP, SMA, SNA and CEP to

---

[1] https://ec.europa.eu/home-affairs/sites/homeaffairs/files/what-we-do/policies/european-agenda-security/20180613_final-report-radicalisation.pdf

[2] http://ec.europa.eu/research/participants/data/ref/h2020/wp/2016_2017/main/h2020-wp1617-security_en.pdf

tackle LEAs needs in terms of prevention and action regarding terrorist social media online activity.

The novelty the project brings is combining these technologies for the first time in an integrated solution that will be validated in the context of five LEAs.

The consortium was designed to gather together all required capabilities and expertise that sustain the development of RED-Alert solution:

Five Law Enforcement Agencies (LEAs): Protection and Guard Service from Republic of Moldova (SPPS), Guardia Civil from Spain (GUCI), Ministry Of Public Security – Israel National Police (MOPS-INP), Metropolitan Police Service from UK (SO15) and Protection and Guard Service from Romania (SPP);

Five Industrial innovation champions (of which four SMEs): SIVECO Romania SA (SIV), Intu-View Ltd (INT), Usatges Bcn 21 Sl (INSKT), Maven Seven Solution Technology (MAV), and Information Catalyst for Enterprise Ltd (ICE);

Four Academic & Research Organizations: Interdisciplinary Center Herzliya (ICT), Eotvos Lorand Tudomanyegyetem (ELTE), City University Of London (CITY) and Birmingham City University (BCU);

One Regulatory association: Malta Information Technology Law Association (MITLA).

The project duration is 36 months and started in June 2017.

9.2 Research Challenges Addressed

The main challenge in the domain of terrorism and radicalization research is that the underlying data sources and data usages are constantly and rapidly evolving, as terrorist groups are moving away from structured written blogs and forum posts and instead, are using social media to propagate URLs that redirect to repositories of propaganda videos. Thus, processes of detecting suspicious content can become quickly outdated, and it is becoming essential to automatically adapt the system to evolving media channels layouts and interfaces, as well as changing user behaviours.

To support the project's objectives, the following key performance indicators (KPIs) are to be reached until the end of the project: seven social media channels mined for content, 10 languages supported for analysis, improved accuracy and usability of tools within the context of data privacy, as well as extended real-time and collaborative capabilities and support for further development.

To address these KPIs, the RED-Alert project mixes relevant software components from different partners. In the same time, the challenge and innovation are to combine technologies such as CEP, SNA and NLP to assess social features in communications used by terrorist organizations. This will imply harmonisation of theories, tools and techniques from cognitive science, communications, computational linguistics, discourse processing, language studies and social psychology. Moreover, in order that the system performance to be adapted for each component the project implements a meta-learning process that will assist SNA, CEP and NLP components defined processes.

Another major challenge that needs to be addressed by the project is to preserve the privacy of citizens that use online social networking platforms. Having in mind the rumours linked with social media data collection and new GDPR that applied from 25th of May 2018, became obvious that the Internet service providers struggle to balance the user privacy against the national security. The only way to move forward is to preserve the privacy when processing the data and in the same time to take advantage of the latest technological advancement when designing the security part of the system; hence, the malicious content and the corresponding personality can be tracked while the privacy of innocent citizens can be preserved. RED-Alert system will include privacy-preserving mechanisms allowing the capture, processing and storage of social media data in accordance with applicable European and national legislations.

RED-Alert will face the additional challenge of allowing collaboration between the different LEAs from different countries, with different privacy laws and trust levels by implement a privacy-preserving tool to mine the data.

There is a growing understanding that innovation, creativity and competitiveness must be approached from a "design-thinking" perspective – namely, a way of viewing the world and overcoming constraints that is at once holistic, interdisciplinary, integrative, innovative, and inspiring. Privacy, too, must be approached from the same design-thinking perspective. Privacy must be incorporated into software systems and technologies, by default, becoming integral to organizational priorities, project objectives, design processes, and planning operations [2].

9.3 Architecture Overview

The vision of RED-Alert project is to develop and validate a real-time
system able to facilitate the timely identification of terrorism-related content
by summarizing large volumes of data from social media and other online
sources (such as blogs, forums).

The RED-Alert components as shown in Figure 9.1: NLP, SMA, SNA,
CEP, Data Anonymization, Data Visualisation and ML) will be integrated
in three separate layers, based on Lambda Architecture[3] concepts defined by
[3], designed to handle massive quantities of data by taking advantage of both
batch and stream methods for real-time data processing, as follows:

- the "speed" layer, which includes data acquisition components for
  processing data streams in real time by means of data collection
  (social media capture, web crawling, LEA "raw" content), data filtering

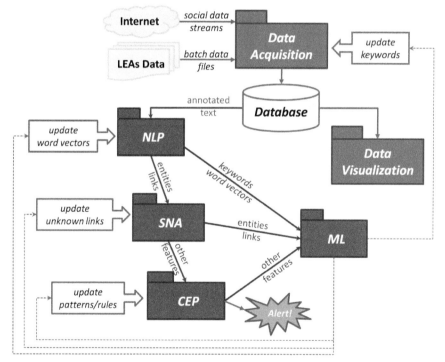

Figure 9.1 Dynamic learning capabilities of the systems to update keywords, vector spaces,
rule patterns, algorithms and models.

---

[3]http://lambda-architecture.net/

(pulling text data from a message queue, normalizing and extracting the required meta-data), data enrichment (multimedia content analysis), and data privacy (anonymization of text and image data);

- the "batch" layer, which integrates the predictive models (based on CEP) that will be used by the pattern detection features within the analysis module. Due to the changing nature of the facts and behaviours, the set of stored models should be periodically re-trained with the new data arriving to the system. This is usually a resource-intensive task that cannot be performed in real-time and it should be scheduled as a frequent batch job;
- and the "service" layer, which integrates the visual analytics gateway that will be in charge of presenting the aggregated data, metrics and events configured by users, who can set up the rules or conditions for triggering alerts. This will be used directly by the rules engine to determine whether the conditions exist for a particular event type. The layer will also offer a Web Service API allowing third parties or LEAs in-house developers to build external components on top of the RED-Alert integrated solution.

Figure 9.2 shows the designed Architecture for RED-Alert. In this multi-layered architecture, application components are grouped into logical layers, namely:

- Front-end Layer – grouping components and functionalities that face the end-users of the system, with the role of getting and presenting data, displaying alerts, and allowing the users to configure the system and administrators to monitor it;
- Back-end Layer – grouping the core modules and data processing components that service the system;
- Integration Layer – grouping inward middleware services that inter-connect the components of the system, as well as outward facing APIs that facilitate connections with other systems;
- Data Storage Layer – grouping database management systems (both relational and non-relational) that handles the storage of data needed by the system.

This approach to architecture, described above, attempts to balance latency, throughput, and fault-tolerance by using batch processing to provide comprehensive and accurate views of historical operational batch data, while simultaneously using real-time stream processing to provide views of online data.

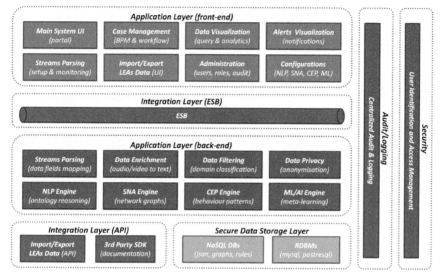

Figure 9.2 Layered application architecture.

The "speed" layer sacrifices throughput as it aims to minimize latency by providing real-time views into the most recent data. The "batch" layer pre-computes results using a distributed processing system that can handle very large quantities of data. Output from the "batch" and "speed" layers are stored in the "service" layer, which responds to ad-hoc queries by returning pre-computed views or building views from the processed data.

"Privacy by Design", focuses on maximizing privacy and data protection by embedding safeguards across the design and development of software systems, services or processes by taking privacy and data protection considerations into account from the outset and throughout their whole lifecycle, rather than as a remedial afterthought. Such safeguards should be built into the core of the products, services or processes and treated as a default setting for not only technologies, but also into used operation systems, network infrastructures, work processes and management structures [4].

9.4 Results

9.4.1 Natural Language Processing Module (NLP)

Ever since the Tower of Babel, the human race has taken recourse to translation to bridge the gap between languages, cultures, societies and nations. Translation serves many purposes: it enables us to broaden the scope of our cultural perspective, to see the world in a way that others – friends and foes – do, to retrieve ancient knowledge that, otherwise, would be lost to mankind and to communicate between people on a day to day basis.

However, in a global environment challenged with enormous amounts of information, a challenge has arisen that cannot be solved by translation. This is the need to identify affinities and dis-affinities between semantic units in different languages to normalize streams of information and mine the "meaning" within them regardless of their original language. When we look for information or wish to generate alerts – particularly in domains that are global – we do not want to be restricted to streams of information in one language; when we are interested in information – be it alerts on terrorism, fraud, cyber attacks or financial developments – we do not care if the origin is in English, French Arabic, Russian or Chinese. The need, therefore, is for technology that scans the entire gamut of information, identify the language and the language register of the texts, perform domain and topic categorization and match the information conveyed in different languages to create normalized data for assessment of the scope and nature of a problem.

The problem facing automated extraction of meaning from language is not restricted to translation between languages but within languages. That which we call a "language" is frequently a political definition and not one based on the linguistic reality. Some cases of a "language" are, actually a group of "dialects" that in other cases are defined as separate languages. The decision to call Swedish, Danish and Norwegian separate languages on one hand, and Moroccan, Libyan, Saudi Arabia and Egyptian all "Arabic" is political and not linguistic. Even within the same language register, words, quotations, idioms or historic references can be "polysemic"; they have different meanings according to the domain and the context of the surrounding text. A verse in the Quran may mean one thing to a moderate or mainstream Muslim and the exact opposite to a radical.

Methods to deal with this problem have generally been based on multilingual dictionaries that enable key words spotting (by input of a key word in one language, the search engine can add the nominal corresponding terms in other languages) or by automated translation of texts and application of the search criteria in the target language. The limitations of such methods are obvious: a word in one language has many "translations" and not all of them may even be remotely related to the meaning that the user is interested in.

In 1949 the cryptologist Warren Weaver wrote a memorandum on automated translation using computer technology. Weaver suggested the analogy, of individuals living in a series of tall closed towers, all erected over a common foundation. When they try to communicate with one another, they shout back and forth, but cannot make the sound penetrate even the nearest towers. But, when an individual goes down his tower, he finds himself in a great open basement, common to all the towers. Here he establishes easy and useful communication with the persons who have also descended from their towers. Thus, he suggested "...to descend, from each language, down to the common base of human communication – the real but as yet undiscovered universal language – and then re-emerge by whatever particular route is convenient" [5]. In this description, Weaver touched – without calling it by name – on the approach that we are suggesting: semantic normalization of statements in different languages according to domain-specific ontologies.

This solution is based on emulation of the "intuitive" links that domain experts find between concatenations of lexical occurrences and appearances of a document and conclusions regarding the authorship, inner meaning and intent of the document. In essence, this approach looks at a document as a holistic entity and deduces from combinations of statements meanings, which may not be apparent from any one statement. These meanings constitute the "hermeneutics" of the text, which is manifest to the initiated (domain specialist or follower of the political stream that the document represents) but is a closed book to the outsider. The crux of this concept is to extract not only the prima facie identification of a word or string of words in a text, but to expand the identification to include implicit context-dependent and culture-dependent information or "hermeneutics" of the text. Thus, a word or quote in a text may "mean" something that even contradicts the ostensible definition of that text.

The meanings that are represented in one language by one word may be represented in other languages by completely different lexemes (words). "Idea Analysis" or "Meaning Mining" is the ability to extract from a text the hermeneutics (interpretation) that is not obvious to the non-initiated reader. We use of "Artificial Intuition" technology for this purpose. Artificial Intuition is based on algorithms that apply to input of unstructured texts the aggregated comprehension by seasoned subject matter experts regarding texts of the same domain used in training. Humans reach "intuitive" conclusions – even by perfunctory reading – regarding the authorship and intent of a given text, subconsciously inferring them from previous experience with similar texts or from extra-linguistic knowledge relevant to the text. After accumulating more information through other features (statements, spelling and references) in the text, they either strengthen their confidence in the initial interpretation or change it. These intuitive conclusions are part of what the Nobel Laureate Prof. Daniel Kahneman called "fast thinking" – a judgment process that operates automatically and quickly, with little or no effort and no sense of voluntary control [6].

We have approached this problem through combining language-specific and language-register specific NLP with domain-specific ontologies. The technology extracts such implicit meaning from a text or the hermeneutics of the text. It employs the relationship between lexical instances in the text and ontology – graph of unique language-independent concepts and entities that defines the precise meaning and features of each element and maps the semantic relationship between them. As a result of these insights, the process of disambiguation of meaning in texts is based on a number of stages:

- Identification of the "register" of the language. The register of the language may represent a certain period of the language), dialects, social strata etc. In the global world today, however, it is not enough to identify languages; the world is replete with "hybrid languages" (e.g. "Spanglish" written and spoken by Hispanics in the US; "Frarabe" written and spoken by people of Lebanese and North African origin in France and Belgium) that are created when a person inserts secondary language into a primary (host) language, transliterates according to his own literacy, accent etc. It is necessary, therefore, to take the non-host language tokens, discover their original language, back transliterate them and then find the ontological representation of that word and insert it back into the semantic map of the document;
- Identification through statistical analysis (based on prior training of tagged documents) of the ontological instances in the text to determine the probability that the author represents a certain background and ideological leaning. Statistical categorization of a document as belonging to a certain domain, topic, or cultural or religious context can reduce the number of possible interpretations of a given lexical occurrence, hence reducing ambiguity;
- Disambiguation using the immediate neighbourhood of the lexical instances. Such neighbourhood consists of the lexical tokens directly preceding or following the lexical instance. After reading a number of texts of a given genre, the algorithm infers that X percent accord to statement A, the meaning B. When statement C is encountered in a text that is categorized as belonging to the same genre, the algorithm derives

from this a high level of confidence that C also means B. This confidence can be enhanced by additional information in the text;

- Statistical categorization of a document as belonging to a certain domain, topic, or cultural or religious context to reduce ambiguity;
- Chunking and Part of Speech Analysis of the text to use the relationship between different words (not necessarily arbitrarily choosing a certain level of N-grams) to provide additional disambiguating information;
- Based on the identification of the domain of the text, the lexical units (words, phrases etc.) are linked to ontological instances with a unique meaning (as opposed to words which may have different meanings in different contexts) that can be "ideas", "actions", "persons" "groups" etc. An idea may be composed of statements in different parts of the document, which come together to signify an ontological instance of that idea[4];
- The ontological digest of the document then is matched with pre-processed statistical models to perform categorization.

This approach, therefore, is not merely "data mining" but "meaning mining". The purpose is to extract meaning from the text and to create a normalized data set that allows us to compare the "meaning" extracted from a text in one language with that, which is extracted from another language.

This methodology applies also to entity extraction. Here, the answer to Juliette's queclarative, "what's in a name" is – quite a lot a not – as Juliette suggested almost nothing. A name can tell us gender, ethnicity, religion, social status, family relationships and even age or generation. To extract the information, however, we must first be able to resolve entities that do not look alike but may be the same entity (e.g. names of entities written in different scripts English, Arabic, Devanagari, Cyrillic) and to disambiguate entities that look the same but may not be (different transliterations of the same name in a non-Latin source language or culturally acceptable permutations of the same name).

---

[4]Ontology is a graph of unique language-independent concepts and entities built by experienced subject matter experts that defines the precise meaning and features of each element in the graph and maps the semantic relationship between them. Hence, the features that are encountered in the surroundings of a lexical instance are factored in the system's decision to what unambiguous meaning (ontological instance) to refer the lexical instance. "Ontology", Tom Gruber, Encyclopedia of Database Systems, Ling Liu and M. Tamer Özsu (Eds.), Springer-Verlag, 2009.

9.4.2 Complex Event Processing Module (CEP)

The key challenges so far with the complex even processing has been the need to make it both functional and generic at the same time. As downstream consumers, the component is dependent on receiving data from the other, upstream components, like the NLP and SMA data. The challenge here was to produce something that could consume unknown data as well as make assumptions and best guesses as to the nature, structure, quantity and quality of the data. In addition to working in the dark with its source data, the CEP engine also had to the challenge of not having any intelligence data to work with either. Clearly, on a project of this nature LEAs must guard and protect their intelligence for a plethora of operational reasons, however, regardless of this the CEP engine must still be delivered and demonstrate a working capability, so again it had to make a few leaps of faith which in the end should remain relevant when integration and pilot tests them out. Hence, the CEP engine remains probably a bit simplistic because of its generic nature, but by the same token generic is inherently extensible – so as both upstream data and real-world intelligence are fed into it, the engine will be able to adapt.

The CEP component aims to identify, via pattern matching algorithms, the dynamics, interactions, feedback loops, causal connections and trends associated with the data content it receives as input from the other RED-Alert components. Specifically, it is a secondary, downstream consumer of pre-processed data from the NLP, SNA and AI components and will generate output alerts; the component will also allow the configuration of data sources to allow the ingestion of external data out with the primary sources. The alerts themselves will be output to log files which will be monitored by a file reader component to display alerts, as well as monitor the CEP engine as a whole, and to integrate with the external API's of the LEAs.

As a development timeline, it comprises template architecture of many different CEP nuances which are set/selected/derived via a web tool to produce myriad applications. A data ingestion component will, either acquire processed data from the configured input component via configurable connection components, or the connection components will feed Kafka topics which will serve as the actual source for all CEP input. Apache Kafka[®5] is a distributed streaming platform generally used for two broad classes of applications: 1) for building real-time streaming data pipelines that reliably

---

[5]https://kafka.apache.org/

get data between systems or applications, and 2) for building real-time streaming applications that transform or react to the streams of data. And it is in this 2nd type of applications where the CEP from RED-Alert is conceived. Current expectations assume that multiple CEP applications will be running in parallel – each either working on different parts of the input data, or on different patterns within the data, or different configurations of the same CEP Application but utilizing an alternate configuration (e.g. on data consumed per month versus per week) or providing staged, partial result sets that will subsequently be consumed by an additional, downstream, CEP application that will act on the staged data.

Other tools and technologies covering similar RED-Alert needs and functionalities were analysed but dismissed as there was a need for extra developments further than the actual accomplished or the expertise of the development team was more limited. These other technologies were Apache Flink which is incorporated into the main CEP RED-Alert component, Spark or Red Hat Drools.

Primarily from a performance perspective, we expect Kafka to deal with this sort of load far better than Mongo, hence we expect any data sourced from Mongo to be moved into Kafka, and hence a data loading component will perform this task. Note also that as part of creating staged, pre-processed data for downstream consumption by other CEP applications, the CEP applications themselves will create and populate MongoDB/Kafka topics as well. Also, it is likely that Kafka will serve as the primary source for the engines and that these topics are populated from MongoDB, in real time. This event data is then converted to a data type associated with the CEP software via a generic parsing component to produce objects with a common structure representative of the source data (i.e. NLP, SNA and ML).

The block diagram shown in Figure 9.3 outlines the workflow, interactions, input/output and decision-making processes on the CEP engine itself. As the diagram clearly shows, the engine itself works on structured, well defined JSON, where well defined includes all field names, their data types as well as an indication of their original source – Note, in this case, source indicates where the data analytics (i.e. NLP, SNA and SMA processing) that generated particular aspects of the JSON originated, as opposed to the source of the input, i.e. the raw data.

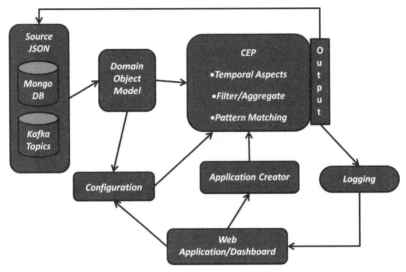

Figure 9.3   Complex event processing module – Logical component diagram.

9.4.3 Semantic Multimedia Analysis Tool (SMA)

Multimedia is extensively used in social networks nowadays and is gaining popularity among the users with the increasing growth in the network capacity, connectivity, and speed. Moreover, affordable prices of data plans, especially mobile data packages, have considerably increased the use of multimedia by different users. This includes terrorists who use social media platforms to promote their ideology and intimidate their adversaries. It is therefore very important to develop automated solutions to semantically analyse given multimedia contents. The SMA Tool is designed to ensure security and policing of online contents by detecting terrorist material.

The SMA Tool extracts meaningful information from multimedia contents taken from social media. The five main features of the tool are:

- Segmentation of audio streams, identifying sections of speech;
- Transcription of the segmented speech sections using an ASR engine;
- Detection of sound events within audio streams, such as gunfire, explosions, crowd noise etc;
- Extraction and identification of objects, such as logos, flags, weapons, faces, etc., within image and video scene elements;
- Extraction and transcription of text elements in image and video elements.

Moreover, the SMA Tool retrieves multimedia data, converts it to a uniform format and delivers the analysis results. The extraction of semantic information is the third of four stages the tool will perform. All four stages are as follows:

- Input: Retrieval of multimedia files from disk or URL;
- Stream Separation: Extraction of audio/video streams in multimedia files;
- Feature Analysis: Semantic analysis of audio/image content;
- Output: Compilation of results in a uniform JSON format.

The results of this tool are sent to the other key components of the project such as NLP, SNA and CEP.

9.4.3.1 Speech recognition

This component is used for audio segmentation, language detection and speech transcription. The RED-Alert project is required to support 10 languages, and be able to run offline, without having to send data to a 3rd party web API. We have consulted our LEA partners to prepare a list of 10 languages which must be supported by the speech/written text transcription elements of the SMA Tool. These languages are: Arabic, English, French, German, Hebrew, Romanian, Russian, Spanish, Turkish, and Ukrainian.

9.4.3.2 Face detection

The SMA tool uses a Haar-like feature based cascade classifier [7] to detect both frontal facing and profile faces in images. Haar-like features are calculated by finding the difference in average pixel intensity between two or more adjacent rectangular regions of an image. In the SMA tool, Haar cascades are used as a supplementary feature to implement simple face detection. More advanced techniques are implemented in the object detection element, which can also be used to detect people/faces.

9.4.3.3 Object detection

State of the art methods for detection of objects within images use large neural networks consisting of multiple sub-networks (region proposal network, classification network etc.). The SMA tool's object detection utility uses the Faster R-CNN structure [8]. Faster R-CNN is constructed primarily of two separate networks: a Region RPN which produces suggestions of regions of an image which might contain objects, and a typical CNN which generates a feature map and classifies the objects in the proposed regions.

9.4.3.4 Audio event detection

Audio event detection is implemented in the SMA Tool by using a recurrent CNN [9]. The convolutional element classifies the short term temporal/spectral features of the audio, while the recurrent element detects longer term temporal changes in the signal. The SMA Tool applies feature extraction prior to processing by the network. This provides a more detailed representation of the audio signal to the network, meaning the first few layers can extract more meaningful information. Peak picking algorithms [10] are applied to remove any noise and only annotate the onset of any detected audio events.

9.4.4 Social Network Analysis Module (SNA)

In the last decades, human communication has gone through a crucial transition. Thanks to the Internet, which connects all individuals around the globe, everybody can contact each other without any time delay and without geographical restrictions. Social interactions became cheap and worldwide, the only restriction remained at the human side: all of us are able to process information at a finite rate and can engage trustful relations only with a few tens or hundreds of others. Therefore, describing and modelling of the new type of human interactions called for a description which is free of space limitations: these represent the tools of Network Science.

SNA module, aims to provide methods and software solutions for handling relational data. It focuses on three aspects of networked analysis as described in the following subsections.

1) *Network dynamics and temporal network structure models.*

The tool describes the evolution of networks and edges/nodes in time, by calculating quantitative features derived from models on evolving networks, and evolution of communities.

Real systems are usually not static, instead they evolve in time [11]. This can manifest in the emergence of new parts, the disappearance of existing parts, and also the relations among constituents can be rearranged over time. Temporal networks with changing topology over time result typically changing community structures. Since community finding methods determine the structures only at different time steps, the structures from consecutive steps must be matched. When communities simply shrink or increase in size, then the matching is straightforward: matching of communities is determined uniquely by intersecting nodes between the two communities of different time steps. However, individuals can also change their community membership over time.

The SNA module implements a special community finder algorithm to solve this challenge. The solution is based on the property of the applied algorithm, which ensures, that adding new nodes and edges to a network does not change the membership status of a node or an edge. The only possible change is, that distinct communities fuse. This property allows an algorithm to match consecutive groups by introducing an intermediate time step, where the two snapshots are merged into a common network. Because the intermediate snapshot can contain only additional nodes and edges, the communities of the intermediate network can be matched to the prior and to the subsequent communities by the rule of matching intersections.

2) *Link prediction solution*

The SNA tool of the RED-Alert solution adopts network theoretic similarity and distance measures for counter terrorism purposes. Based on the special targeted measures, missing links and nodes are predicted by the module. Furthermore, some features e.g. weights, labels, directionality of the links are updated as well.

The implementation relies on two theoretic pillars:

- prediction based on topological measures;
- prediction based on attribute information.

Topological measures use only information from connectivity patterns, in contrast attribute measures predict missing/hidden relations from common attribute statistics. Upon request of the analyst on the user interface of the integrated RED-Alert solution, the SNA module can apply hybrid predictions as well, where both networked measures and attribute data are combined (Soundarajan & Hopcroft, 2012).

It must be noted though, that all theoretical speculations are useless without reliable data sources. The scientific background behind this tool ensures only the mathematical rigour with the calculations, but the final conclusions must be always thoroughly reviewed by human experts. All mathematical models work with assumptions that can be only partially valid in real scenarios.

3) *Hierarchy reconstructing methods*

Terrorism has its own frame and structure. As all organizations that consist of many individuals and conduct several tasks, the actions of terrorists are driven by a hierarchical background. However, in several cases, this hierarchy is hidden and builds up in a self-organized way. For traditional observation techniques, this organization seems to be wide spread, unstructured and loosely connected. Here comes SNA into an important role: collecting small

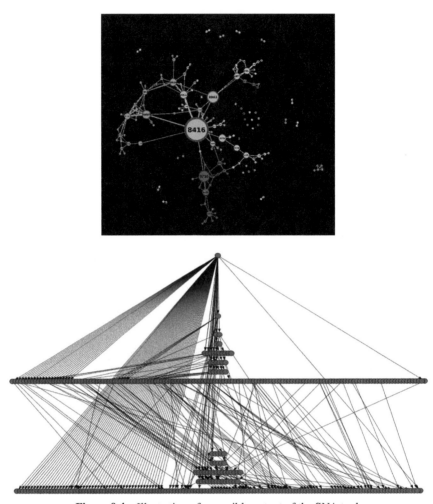

Figure 9.4   Illustration of a possible output of the SNA tool.

pieces of information from huge amount of data results in a holistic picture, where – if data allows it – the unseen hierarchical skeleton can be revealed.

Here, algorithms are implemented for revealing hierarchical structures from flat dataset. New networks are constructed from input data: either from co-occurrence statistics or from directed networks containing loops. Furthermore, quantitative measures are calculated for characterizing the similarity of any network to an ideal hierarchical structure [12].

The upper drawing in Figure 9.4 presents a typical thread-network layout of a forum in the Darkweb. The thread IDs are shown within nodes and the size of the nodes is proportional to the edges belonging to the given node. Node colours indicate topic groups; links are coloured by the dominant neighbouring node. The lower drawing in Figure 9.4 shows the hierarchical structure of commenters of a Darkweb forum.

9.5 Data Anonymization Tool

We live in an era of technology, where smart devices surround us in all realms of life. These devices feed on our information to generate smart options for us, which at the end help us in making smart decisions. The data gathered by these devices can contain vital personal information such as name, age, location and interest. Alongside these smart devices, nowadays, we tend to rely on social network to broaden the scope of our social interactions. We share personal information such as name and age, we highlight the key things happening in our lives such as places visited, accidents and achievement, we also like sharing our believes and interests. Unlike smart devices where adversaries need to corroborate with others to gather information about a single individual, social network data is a source of detailed insight into one's life, thus becoming a bigger threat compared to a single smart device. To mitigate the potential risks, the General Data Protection Regulation (GDPR) was introduced. This new regulation limits the way in which personal data is processed. It limits the ways in which data processing can be done by providing only six lawful ways: Consent, Contract, Legal obligation, vital interest, Public task and legitimate interest[6]. In light of this new regulation, processing social network data becomes tricky. Data collectors, which own the social networks can process this information but after explicitly informing the data owner. This limits the flexibility that third party organizations had. Under the GDPR, all third party companies, who do not have prior consent, need to rely on anonymized data only.

Data anonymization has been around for a while now. It is a process of carefully categorizing social network data into different streams, where each stream undergoes a certain set of tasks. Social network can be divided into three main streams, personal identifiers, quasi-identifiers and non-personal data. Personal identifiers refer to all such parameters that can help identify an individual directly from a large dataset. This mainly constitutes of name, unique-id, contact number and email address. To ensure data anonymization, all such data is removed from the dataset before further processing, thus

---

[6]https://ico.org.uk/for-organisations/guide-to-the-general-data-protection-regulation-gdpr/lawful-basis-for-processing/

reducing the probability of identification of an individual from a large dataset. This probability is further reduced by processing the quasi-identifiers, which on their own have limited meaning but when combined with other quasi-identifiers, can lead to privacy violation. For instance, a dataset containing age information would have less meaning, but when combined with location information would help adversaries in narrowing down their search for an individual and the more quasi-identifiers one has the higher the probability of identification. Therefore, quasi-identifiers are key parameter that all anonymization techniques need to process. The third stream of data deals with the non-personal data. This is set of information that is not connected to any particular individual and can point to anyone in the dataset, for instance, a Facebook post or a twitter tweet can be made by anyone thus, this is considered as non-personal data.

To introduce data anonymization, data analyst carefully analyses the dataset and then narrow down the anonymization approaches that need to be executed. Social network contains three types of quasi-identifiers: numeric, non-numeric and relational information. This therefore means that three separate streams of data anonymization techniques are combined to get results for social network data. Numeric data can be handled by the well-known differential privacy approach [13], where as non-numeric data is handled by k-anonymity (Sweeney, 2002). The relational information is anonymized using a privacy conscious node-grouping algorithm [14]. This anonymized data ensures that no individual can be identified from the processed social network data.

The anonymization techniques applied in this project work in hiding information about all innocent individuals but it also helps terrorist organizations in hiding behind the covers. This as a result puts extra burden on the SNA, CEP, NLP modules. They adapt to working on anonymized social network data and narrow down the search of terrorist organizations. Once identified, the LEAs need to know the identity of the highlighted individuals. To cater for this need, a de-identification approach is also developed in this project, that takes as input the surrogate id's that are provided by the anonymization technique and provide the true identity of an individual. This de-identification algorithm only exists due to the nature of the project and where one can argue that this would make the anonymization algorithm pseudo in nature, it is key to highlight that the de-identification approach only resides with the LEAs thus limiting any adversary from actually identifying individuals and also complying with the GDPR.

9.6 Data Networked Privacy Tool

Intelligence information can be very tricky at times and the nature of this information limits LEAs located in different geographical location from sharing information. On the contrary social networks have no territorial boundaries and terrorist organization can operate from any possible location, making it harder for LEAs to track and tackle them. To overcome this difficulty, the Red-Alert project is equipped with a novel Inter-LEA search algorithm. It limits and controls the amount of information that LEAs located in different geographical location can share with the use of high end encryption algorithm. Under this approach, as shown in Figure 9.5, LEAs are independent in performing their own search and collecting their own intelligence information, they then are requested to populate a list of names of the individuals identified. The second LEA who is looking for a particular individual can search in the encrypted list and find out if one exists or not. The benefit of using high end encryption techniques is of limiting what else an inquiring LEA can see. As such, the LEA inquiring only sees a response in terms of a YES or a NO, therefore hiding all other names in the database. The search query is made with taking a probability attack into consideration,

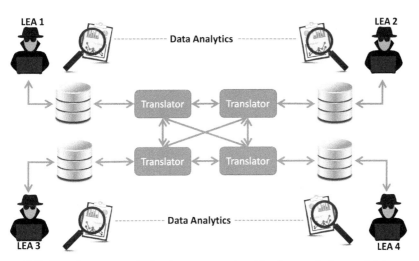

Figure 9.5 Two-layer networked privacy preserving big data analytics model between coalition forces.

thus if an LEA searches for the same name over and over again, there exists no defined pattern. This limits the first LEA (who is hosting the list) from knowing what name is being searched, thus making it a double sided blinded process.

9.7 Integration Component

All the components presented in previous sections will be integrated in one unique solution. The integration component will have therefore some different subcomponents

- **Main System User Interface** provides common look-and-feel to the graphical user interface of the overall RED-Alert System. As shown in Figure 9.6, this component will provide a portal-like user interface for the overall system with common interface placeholders, such as header and footer, main menu, and user interface components hosting through custom common APIs.

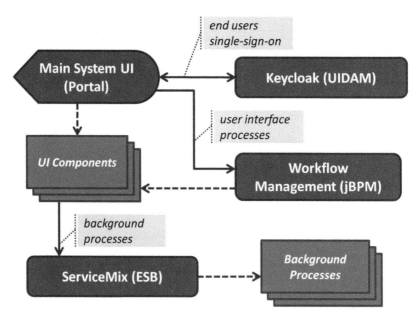

Figure 9.6 Main system user interface – Component interactions diagram.

- **User Identification** and Access Management component will be implemented based on RedHat Keycloak[7] and will provide the means for identifying users and managing their access to application components, both to front-end user interface and to back-end processes;

- **The Collaborative Workflow**/Case Management component is based on RedHat jBPM,[8] a light-weight and extensible workflow engine, offering process management features and tools for both business users and developers. RedHat jBPM supports adaptive and dynamic processes that require flexibility to model complex, real-life situations that cannot easily be described using a rigid process;

- **Application Integration Services** component is built with Apache ServiceMix,[9] an open-source integration container that unifies the functionalities of Apache ActiveMQ, Camel, CXF, and Karaf into a powerful runtime platform you can use to build your own integrations solutions. It provides a complete, flexible, enterprise ready ESB exclusively powered by OSGi;

- **System Interoperability Services** component will be built on top of the Application Integration Services, exposing selected RED-Alert system's functionalities to external system, including existing systems of LEAs;

- **Centralized Audit and Logging** component will be implemented using Audit4j,[10] an open source auditing framework which is a full stack application auditing and logging solution for Java enterprise applications, tested on a common distributions of Linux, Windows and Mac OS, designed to run with minimum configurations, yet providing various options for customization.

Figure 9.7 presents the interactions of the Centralized Audit and Logging component with the Main System UI (Portal), by means of hosting the visual part exposed by the component, and also with the other components of the RED-Alert system, by means of custom common APIs that will allow all components to log entries into a central repository.

---

[7] http://www.keycloak.org

[8] https://www.jbpm.org

[9] http://servicemix.apache.org

[10] http://audit4j.org

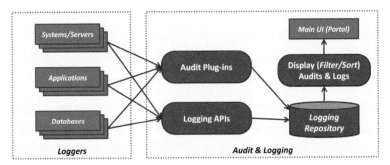

Figure 9.7 Centralized audit and logging – Interactions diagram.

9.8 Future Research Challenges

The next stage of the CEP process is to integrate the recently published standard JSON to finalise the ingestion process (Mongo data source to Kafka topic transfer), and to extend the range of CEP engines to accommodate the SNA (network analysis) data.

One of the major challenges of multimedia extraction is to reduce the number of false positives. We need to make fine grained tuning of SMA tool's components by using larger dataset of a broad range of objects and audio variations. Nowadays data collection, processing and storage have become itself very challenging due to the recently enforced GDPR compliance requirements. The situation is improving with the development of new data management processes and good practices for the data protection. We aim to further improve the performance of SMA Tool and evolve it towards a comprehensive Multimedia Forensics Analysis Toolkit.

Social network analysis is very sensitive to the quality of the available datasets. Further research will aim to develop algorithms for evaluating noisy or biased input datasets. E.g. ensemble averages over possible realizations of networks can shed light on the reliability of predictions.

Another challenge to be addressed is to develop tools for hierarchical visualization of time evolving networks, which helps the analyst in understanding the possible correlations and trends at different scales.

Integration activities will continue with the scheduled iterations towards the piloting phase. These iterations imply adding online streaming capabilities to data acquisition component and expanding the social media channels capabilities beyond Facebook and Twitter, to reach the 7 channels KPI. The Common Schema will be extended with new fields to support these new channels. The security and audit solutions will also be rolled out to all other components to enable the full scope of security requirements of LEAs.

Acknowledgement

This project has received funding from the European Union's Horizon 2020 Research and Innovation Programme under Grant Agreement No. 740688.

References

[1] Migration and Home Affairs, "High-Level Commission Expert Group on Radicalisation (HLCEG-R)," Publications Office of the European Union, Luxembourg, 2018.

[2] OASIS, Privacy by Design Documentation for Software Engineers Version 1.0, 2014.

[3] N. Marz and J. Warren, Big Data – Principles and best practices of scalable realtime data systems, Manning, 2015.

[4] Deloitte, Privacy by Design Setting a new standard for privacy certification, Deloitte Design Studio, 2016.

[5] The Rockefeller Foundation, "Reproduction of Weaver's memorandum," 15 07 1949. [Online]. Available: http://www.mt-archive.info/Weaver-1949.pdf. [Accessed 12 December 2018].

[6] D. Kahneman, "Thinking, Fast and Slow by Daniel Kahneman," *Journal of Social, Evolutionary, and Cultural Psychology,* vol. 2, pp. 253–256, 2012.

[7] P. Viola and M. Jones, "Rapid Object Detection using a Boosted Cascade of Simple Features," in *Conference on computer vision and pattern recognition*, 2001.

[8] S. Ren, K. He, R. Girshick and J. Sun, "Faster R-CNN: Towards Real-Time Object Detection with Region Proposal Networks," in *Advances in Neural Information Processing Systems 28 (NIPS 2015)*, 2016.

[9] I. Goodfellow, Y. Bengio and A. Courville, "Deep Learning," MIT Press, 2017.

[10] C. Southall, R. Stables and J. Hockman, "Improvingpeak-picking using multiple time-steploss functions," in *Proceedings of the 19th International Society for Music Information Retrieval Conference (ISMIR)*, Birmingham, 2018.

[11] N. Masuda and R. Lambiotte, A Guide to Temporal Networks, Singapore: World Scientific, 2016.

[12] E. Mones, L. Vicsek and T. Vicsek, "Hierarchy measure for complex networks," *PLoS ONE,* 2012.

[13] C. Dwork, "Differential privacy: A survey of results. In International Conference on Theory and Applications of Models of Computation," *Springer, Berlin, Heidelberg.,* pp. 1–19, April 2008.

[14] W. Asif, I. G. Ray, S. Tahir and R. Muttukrishnan, "Privacy-preserving Anonymization with Restricted Search (PARS) on Social Network Data for Criminal Investigations," 2018.

10

TRUESSEC Trustworthiness Label Recommendations

Danny S. Guamán[1,7], Manel Medina[2,3], Pablo López-Aguilar[3], Hristina Veljanova[4], José M. del Álamo[1], Valentin Gibello[6], Martin Griesbacher[4] and Ali Anjomshoaa[5]

[1]Universidad Politécnica de Madrid, Departamento de Ingeniería de Sistemas Telemáticos, 28040, Madrid, Spain
[2]Universitat Politécnica de Catalunya, esCERT-inLab, 08034, Barcelona, Spain
[3]APWG European Union Foundation, Research and Development, 08012, Barcelona, Spain
[4]University of Graz, Institute of Philosophy and Institute of Sociology, 8010, Graz, Austria
[5]Digital Catapult, Research and Development, NW1 2RA, London, United Kingdom
[6]University of Lille, CERAPS – Faculty of Law, 59000, Lille, France
[7]Escuela Politécnica Nacional, Departamento de Electrónica, Telecomunicaciones y Redes de Información, 170525, Quito, Ecuador
E-mail: ds.guaman@dit.upm.es; medina@ac.upc.edu; pablo.lopezaguilar@apwg.eu; hristina.veljanova@uni-graz.at; jm.delalamo@upm.es; valentin.gibello@univ-lille.fr; m.griesbacher@uni-graz.at; ali.anjomshoaa@ktn-uk.org

The main goal of TRUESSEC project is to foster trust and confidence in new and emerging ICT products and services throughout Europe by encouraging the use of assurance and certification processes that consider multidisciplinary aspects such as sociocultural, legal, ethical, technological and business while paying due attention to the protection of Human Rights.

TRUESSEC's central recommendation to the European Commission (EC) is a label scheme that can suitably address found issues that is worth developing and testing. While actual software development is beyond the current scope of TRUESSEC, the remainder of this paper describes the characteristics of such a solution, allowing the EC to commission a working prototype should it wish to do so.

At the heart of the proposed solution is a set of prioritized survey questions that take into account a set of core areas of trustworthiness to produce both a visual "transparency" statement that is easy for the citizen to understand, and additionally provides a specific piece of code to enable machine-to-machine integration based on the policy settings of 3rd party users. In this regard, the Creative Commons licensing model[1] is analogous to our proposed solution.

10.1 Introduction

This paper provides a recommendation for a TRUESSEC labelling solution, aimed to show users the level of trustworthiness of applications and services, according to multi factor criteria.

The central task of the TRUESSEC project is to apply an interdisciplinary approach, encompassing ethics, sociology, law and technical engineering, to make recommendations to the European Commission for a certification and labelling of ICT products and services that will foster trust among citizens that use them.

Both the core areas that constitute "trust" (which spans cybersecurity through to branding and user experience), and the various potential fields of application (from web services to cyber-physical systems) means that the remit of this project is very broad indeed.

Nevertheless, the project team values this approach and, as background to our recommendation, has noted that good progress has been made with European legislation which, over time, is likely to enhance levels of citizen trust. Even though the Digital Single Market legal framework is still a work in progress, these advances have resulted in a strong legal foundation to protect the rights of EU citizens entrenched in the Charter of Fundamental Rights of the EU [1].

In addition, pan-European bodies, such as ENISA[2], are progressing well with security and privacy certification and codes of conduct in relatively

---

[1]https://creativecommons.org

[2]See the Proposal for a REGULATION OF THE EUROPEAN PARLIAMENT AND OF THE COUNCIL on ENISA, the "EU Cybersecurity Agency", and repealing Regulation (EU) 526/2013, and on Information and Communication Technology cybersecurity certification ("Cybersecurity Act"), COM/2017/0477 final - 2017/0225 (COD).

new areas, such as security in the Cloud and Internet of Things – although certification remains a voluntary responsibility of the online service providers with little legal implications.

Our research started "evaluating existing trustworthiness seals and labels" [2], and the analysis of these existing schemes showed a general lack of adoption and awareness, as well as poor transparency regarding what is being certified and under what conditions. In fact, citizens tend to employ other indicators of trust (3rd party payment systems, branding, user experience, and user-based recommendation engines) to make decisions about their use of a service, despite how little guarantees they actually offer.

TRUESSEC's research work also went beyond current business practices, technology and legislation to explore the social and ethical questions behind what constitutes trust from users. This is summarized by our criteria catalogue, which was published as deliverable [3].

Given these inputs, there are a number of issues with existing label schemes:

- There are too many labels to provide a common understanding for citizens or service providers
- Businesses tend not to understand the cost/benefits of using labelling
- They are not sufficiently flexible and updated to acknowledge relatively new legislation, such as the GDPR
- They are not inclusive enough to incorporate additional 3rd party certification
- They do not "go beyond the law" to enable service providers to demonstrate that they have taken an ethical, responsible and transparent approach
- They rarely encompass all major components of trust such as safety or "security by design", personal data protection and consumer rights enforcement
- They provide insufficient information on how they are awarded and on the safeguards offered

These shortcomings mean that current labels are often out of date, removed from best practices, poorly understood and therefore little known and used.

10.2 Interdisciplinary Requirements

TRUESSEC.eu Core Areas of trustworthiness are based on the findings from five support studies, **considering the European values and**

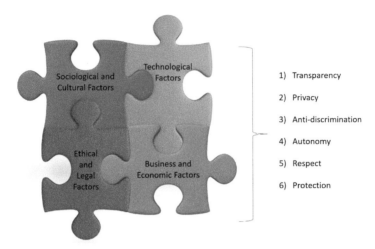

Figure 10.1   TRUESSEC.eu core areas of trustworthiness.

fundamental rights as well as following joint work among all disciplines represented in the TRUESSEC.eu project. Six Core Areas have been agreed upon, that set the stage for the search of the multidisciplinary criteria [4]. The Core Areas displayed in Figure 10.1 represent the reflections of the five disciplines: ethics, law, sociology, business and technology.

Transparency. The TRUESSEC.eu Core Area transparency reflects the understandings of the five disciplines by having information in its focus. In this regard, the Core Area transparency evolves around the fulfilment of information duties related to personal data processing, but it also goes beyond that, as the business perspective shows. Overall, transparency can help to narrow down the existing informational gap and give users clearer answers to questions regarding their personal data and the products and services they purchase.

Privacy. The TRUESSEC.eu Core Area privacy is equally important in all disciplines. When users are provided with relevant information, this sets the ground for them to take control over their data. On the one hand, users must be able to make decisions regarding their personal data; on the other hand, providers must respect those decisions. The latter is a striking point, as providers have commercial interests in processing as many data as possible. Considering the economic relevance of data and the emerging data economy, it is crucial to ensure the protection of personal data. This includes considering aspects of privacy throughout the design and development of an

ICT product or service (privacy by design) as well as offering the privacy settings at a high level of privacy protection (privacy by default).

Anti-discrimination. This Core Area has a great relevance for trustworthiness. The need to formulate such a core area stems from the fact that discrimination concerning ICT products and services is present and it is very often hidden in decision-making carried out by algorithms and self-learning systems. This particularly relates to cases where parameters are included in the decision-making process, which go beyond the scope of the service or product in question.

Autonomy. The TRUESSEC.eu Core Area autonomy summarizes well the considerations of the five disciplines. Having access to and rights to use various ICT products and services brings up one very central issue, which is, the need for users to be given the opportunity to make decisions regarding their personal data. These decisions need to be well informed and free of manipulation and coercion.

Respect. The TRUESSEC.eu Core Area respect presents a transition from discipline-related understanding to a transdisciplinary one. It embodies the idea that based on societal, legal and ethical frameworks there are certain duties that arise for ICT providers that ground legitimate expectations on the side of users when dealing with ICT products and services. Legitimate expectations have three main hallmarks: they are predictive, prescriptive and justifiable. In the ICT context, this would suggest that users create expectations on what ICT providers will and should or should not do, or how they will and should operate. Whereby these expectations are justifiable, that is, users have justification or warrant for forming them in the first place. Example of such legitimate expectations is that ICT providers respect users' rights and freedoms.

Protection. The considerations of all five disciplines seem to be focused in the protection of individuals against any harms as well as the protection of their rights and freedoms. This has led us to formulate the TRUESSEC.eu Core Area protection as the sixth core area. In the context of ICT, protection relates to both safety and security thus encompassing risks of physical injury or damage and risks related to data such as unauthorized access, identity theft etc. In order to enable solid level of protection, compliance with already established safety and cybersecurity standards is essential. The aim is to hinder any harms that may be caused because of using ICT in the first place.

10.3 Criteria Catalogue and Indicators[3]

The TRUESSEC.eu Criteria Catalogue represents a constituent part of the TRUESSEC.eu work on labelling. It is a multidisciplinary endeavour to compile a list of criteria and indicators that could contribute towards enhancing the trustworthiness of ICT products and services. The development of the TRUESSEC.eu Criteria Catalogue consists of two phases: (a) development of the First Draft Criteria Catalogue, which includes only ethical and legal criteria and indicators, and (b) development of the multidisciplinary Criteria Catalogue, which builds upon the First Draft Criteria Catalogue, but it also includes sociological, business and technical input.

The basis for the Criteria Catalogue consists of the European values as stated in Article 2 of the Treaty of the European Union and the European fundamental rights, on the one hand, and the findings from the five support studies prepared in the first year of the project as well as some interdisciplinary work and discussion, on the other hand. It is from here that we extracted the hierarchical structure of the Criteria Catalogue. As depicted in Figure 10.2, we started with high-level concepts we called **Core Areas**. The very aim of the Core Areas is to provide a framework which in a next step could be broken down into elements that are more specific. In that sense, the Core Areas reflect the values that should be considered in the design and use of ICT products and services, and thus serve as an orientation tool when determining the criteria. Based on the Core Areas we then developed the **criteria**. The criteria show what requirements an ICT product and service should fulfil in order to be considered trustworthy. In the hierarchical structure, the criteria are less abstract than the Core Areas; however, they are still not concrete enough to be measurable. For that purpose, we formulated **indicators,** which could be measured. A set of indicators is determined for each single criterion. The aim of the indicators is to indicate the degree to which a particular criterion is met.

Based on the support studies and the interdisciplinary discussion we defined six TRUESSEC.eu Core Areas of trustworthiness: *transparency, privacy, anti-discrimination, autonomy, respect* and *protection* and provided a TRUESSEC.eu multidisciplinary understanding of each of them (see Table 10.1).

---

[3]For more on the TRUESSEC.eu Criteria Catalogue see Stelzer et al. "TRUESSEC.eu Deliverable D7.2: Cybersecurity and privacy Criteria Catalogue for assurance and certification," 2018, https://truessec.eu/library .

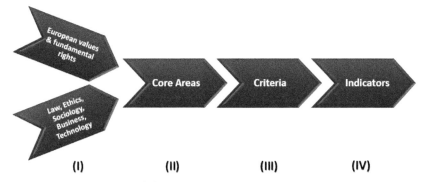

Figure 10.2 Developing the criteria catalogue.

Table 10.1 TRUESSEC.eu Core Areas of trustworthiness

TRUESSEC.eu Core Areas	Multidisciplinary TRUESSEC.eu Understanding
Transparency	The ICT product or service is provided in line with information duties regarding personal data processing and the product/service itself.
Privacy	The ICT product or service allows the user to control access to and use of their personal information and it respects the protection of personal data.
Anti-discrimination	The ICT product or service does not include any discriminative practices and biases.
Autonomy	The ICT product or service gives users the opportunity to make decisions and respects those decisions. The ICT product or service also respects other parties'/persons' rights and freedoms.
Respect	ICT products or services are to be provided in accordance with the legitimate expectations related to them.
Protection	ICT products and services are provided in accordance with safety and cybersecurity standards.

To give a better understanding of the interdisciplinary nature of the Core Areas, we show some exemplary details on the discussion of transparency. From an **ethics** perspective, transparency relates to two aspects: (a) providing clear and sufficient information about products and services in general and (b) more specifically providing information to users regarding activities with their personal data. **Legally**, transparency can be understood as in information duties, laid down in the GDPR, the Directive on consumer rights or the e-commerce Directive. With respect to personal data, transparency is one of the core principles of data processing (Article 5 GDPR). From a **technology** perspective, transparency (in data protection) is defined as the property that all personal data processing can be understood (intelligible and meaningful) at any time by end-users (i.e., before, during, and after processing takes place). In the technical domain there is also a concept named 'Service Level Agreement', which describes technical specification of the service/product being used. You may think e.g. on a service availability, uptime, etc. These more normatively oriented definitions can also be complemented by a **sociological** perspective, which focusses on public opinion. Considering that currently (Eurobarometer data from 2015):

(a) only a minority of EU citizens reads privacy statements (less than 1/5),
(b) only about 4 out of 10 of internet users read the terms and conditions on online platforms,
(c) over 90 % want to be informed if their data ever was lost or stolen,

It can be assumed that there is a need for improvement in current information practices.

Having well-informed citizens, e.g. on the risks of cybercrime, also leads to improved cybersecurity behaviour, which emphasizes the importance of transparency and information. These interdisciplinary considerations can also be connected to a **business** perspective. Transparency includes a wide range of business processes which range from being clear about terms of use of the online service, through to publishing transparency reports about the passing on of user data to 3rd parties, such as law enforcement. Transparency of service and use of personal data is increasingly being perceived by business as a competitive advantage.

From the six Core Areas we extracted the following twelve criteria of trustworthiness:

- Information
- User-friendly consent
- Enhanced control mechanisms
- Privacy commitment
- Unlinkability
- Transparent processing of personal data

- Anti-discrimination
- Cyber security
- Product safety
- Law enforcement declaration
- Appropriate dispute resolution
- Protection of minors

It should be emphasized that the way the criteria are ordered in this list does not indicate their importance per se. Furthermore, we consider this list of twelve criteria to be the groundwork consisting of the most fundamental criteria in the context of ICT products and services. In that sense, the list is not complete from the simple reason that with the technological developments additional criteria might have to be added.

In what follows, we will choose one criterion from the list and use it as an example to elaborate our approach. Table 10.2 illustrates this example.

The Criteria Catalogue is represented in a tabular form. It consists of three columns. The middle column represents the criterion. The right column represents the indicators. As the table shows, to each criterion a set of corresponding indicators are assigned that, when checked, should show to what degree the criterion is fulfilled. In the column on the left, which is named '**Trustworthiness enhancer**', are represented the six Core Areas into six sections. By adding this column, we wanted to show the interrelation between the criterion in question and the Core Areas. In order to show this, we used a colour system. We divided each of the six sections representing the six Core Areas into three subsections where a colour can be applied that would indicate the degree to which based on our assessments the criterion addresses each Core Area. In that sense, one could apply colour to one, two

Table 10.2 Criterion – Information

TRUSTWORTHINESS ENHANCER				CRITERION	INDICATORS
Transparency				Information	i. Information is provided:
Privacy					a. In a user-friendly manner
					• In a plain language (understandable to lay persons)
					• As long as necessary and as short as possible (e.g. in a form of one pager)
Anti-discrimination					b. Relevant to the context
Autonomy					c. Clearly visible and easy to locate
					d. In a structured machine-readable format.
Respect					ii. Information is provided free of charge.
Protection					

or three boxes, with three meaning the criterion fully addresses and meets the particular Core Area. This proved to be, eventually, a very useful way to check whether the group of criteria we identified sufficiently addresses the identified six Core Areas [5].

In this is represented the criterion 'Information'. Our findings showed that information plays undoubtedly an important part in enhancing trustworthiness of ICT products and services. Having the relevant information allows one to make informed decisions and it also creates a climate of openness, and transparency. In general, information consists of two aspects:

(a) **content**, namely, *what* the user is informed about, and
(b) **form**, or *how* information is provided.

Since the first aspect, which is related to the content reappears as an indicator in few other criteria, we have not included it here. In that sense, this criterion was limited only to the *form* of the information provided to the user. As the table shows, the indicators we assigned to this criterion should check whether the information is provided in a user-friendly manner, which means that the information is provided in a plain language that is easily understandable also for laypersons, and that it is as long as necessary and as short as possible. Regarding the length, we suggested that information should be provided in a form of one pager. Additionally, the information should be relevant to the context, easy to locate by the user and it should be provided in a structured machine-readable format. Apart from the format, we also included here another indicator which should check whether the information is provided free of charge. This is just one example of how the Criteria Catalogue operated. The same logic was followed for the other eleven criteria.

One of the main features of the Criteria Catalogue is that it adopts a post-compliance or beyond compliance framework. This framework is very similar to the framework suggested by Luciano Floridi [6]. When analysing the Digital, Floridi distinguishes between hard and soft ethics. Hard ethics is, as he explains, *"what we usually have in mind when discussing values, rights, duties and responsibilities–or, more broadly, what is morally right or wrong and what ought or ought not to be done"* [6]. Soft ethics, on the other hand, is post-compliance ethics as it goes beyond the compliance level and hence beyond existing regulation. In that sense, the aim of the Criteria Catalogue is to address this post-compliance or beyond compliance, for the simple reason that compliance is a very important part in making sure that a business acts within the legal framework. Nevertheless, for enhancing trustworthiness and strengthening trust, which is the main focus of the TRUESSEC.eu project, that might not always be sufficient. With this in mind, in the Criteria Catalogue we provide Core Areas, criteria and indicators as possible ways to address the post-compliance level.

The development of the Criteria Catalogue also paved the way for the drafting of the TRUESSEC.eu recommendations.

10.4 Operationalization of the TRUESSEC.eu Core Areas of Trustworthiness

Using Core Areas of trustworthiness as a starting point, a potential set of ICT system properties and detailed operational requirements have been defined. They attempt to bring Social Science and Humanities requirements closer to the technical domain and analyse which of them have already covered by the state-of-the-art and which need more attention from stakeholders. ICT system properties are quality or behavioural characteristics of a system that, ideally, can be distinguished qualitatively or quantitatively by some assessment method. There are several ICT system properties already defined and studied in the technical realm (e.g. security and safety), so the knowledge base around them can be leveraged to analyse and identify the specific operational requirements that need to be met and assessed for a specific ICT product or service. Figure 10.3 provides an overview of how we have mapped the Core Areas (and criteria) in ICT system properties (details can be found in [7]).

Once identified the ICT system attributes, they can become the basis for carrying out an operationalization process and deriving a set of specific operational requirements that can be realised and assessed.

As depicted in Figure 10.4, operational requirements are requirements of capabilities that should be guaranteed by an ICT product or service to satisfy one or more of the aforementioned ICT system properties. Moreover, they can be used as a precursor to the selection of more specific measures or countermeasures that are known as controls. Controls can be of technical nature (i.e. functionality in hardware, software, and firmware), organizational nature (i.e. organizational procedures related to the system environment and people using it), or physical nature (i.e. physical protective devices).

Finally, controls are instantiated using one or more specific techniques, which are found adequate to fulfil requirements of controls.

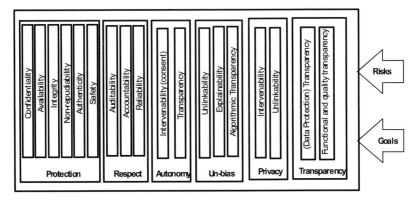

Figure 10.3 Core areas of trustworthiness and related ICT system properties.

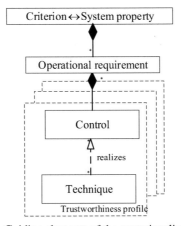

Figure 10.4 Guiding elements of the operationalization process.

It is worth noting the difference between operational requirements, controls, and techniques. Actually, both operational requirements and controls specify a system or organizational capabilities; however, an operational requirement recognises that a trustworthy capability seldom derives from a single control. In other words, one capability, depending on the context, may require several controls. On the other hand, while controls express what measure should be implemented, techniques indicate how it is implemented. Finally, it is important to mention that controls and techniques are context dependent, i.e. they are suitable for the specific context where a system is intended to work. Table 10.3 shows an example of the guiding elements of the operationalization process. Controls and a survey of the technical solutions for trustworthiness can be found in [8].

The state of the practice already includes plenty of controls contained within standard frameworks that, given the broad use of them during audits and certifications, enable to be closer to measurable (and assessable) factors and their corresponding evidence. Controls are widely used by the industry and the state of the practice shows hundreds of standards and certification schemes (around 290 according to ECSO[4]). Just to mention a few examples, security control frameworks include the ISO/IEC 15408 Common Criteria that contains a general catalogue of security requirements for ICT products, the ISO/IEC 27002 defines a set of organizational and technical controls intended to information security management, and the CSA Cloud Control Matrix (CCM) presents a catalogue of cloud-specific security controls. Privacy controls are defined, e.g. in the recent standard ISO/IEC 27018 that is intended to Cloud Service Providers (CSP) acting as Data.

Processors, the NIST 800-53 Rev4 contains security and privacy controls meant to Information Systems and Organizations, and; the General Accepted Privacy Principles (GAPP). Safety requirements e.g. are defined in the IEC 61508-2, they are intended to electrical, electronic, and programmable safety-related systems. Similarly, in the literature we can find significant works that propose, e.g. taxonomies of requirements that can be leveraged to operationalize some of the ICT system properties defined in the section above (e.g. using a goal-oriented approach). For instance, intervenability property can be refined into two guidelines: Data Subject Intervention and Authority Intervention. The first one representing intervention actions for data subjects and the latter the intervention actions for supervisory authorities to intervene in the processing of personal data. Each guideline can be refined into one or more operational requirements that act as success criteria, being empirically observable and objectively measurable. Following up with the intervenability property, the possible intervention actions by data subjects (e.g. do not consent, withdraw consent, review, challenge accuracy, challenge completeness, and request data copy) and the required ICT systems capabilities (e.g. access, no processing, restricted processing, amendment, correction, erasure, data copy, and suspended data flow) may lead to the definition of specific intervention readiness operational requirements. For example, before

---

[4]European Cyber Security Certification, A Meta-Scheme Approach v1.0. December 2017. Available under: http://www.ecs-org.eu/documents/uploads/european-cyber-security-certification-a-meta -scheme-approach.pdf

Table 10.3   Example of the guiding elements of the operationalization process

Guiding Element	Example
Core Area/system property	Protection/Security (Authenticity)
Technical requirement	The system shall provide two-factor authentication for remote access by individuals.
Controls	The system implements multifactor authentication for network access to privileged accounts (NIST 800-53 R4 IA-2-1). The system implements multifactor authentication for network access to non-privileged accounts (NIST 800-53 R4 IA-2-2). The system implements cryptographic mechanisms during transmission (NIST 800-83 R4 SC-8-1).
Techniques	For NIST 800-83 R4 IA-2-1 and IA-2-2, a combination of the following authentication factors can be used: • *Something the principal knows,* such as a password, a personal identification number (PIN), a graphical password, and answers to a prearranged set of questions. A password can be either static or dynamic (e.g. One-Time-password). • *Something the principal has*, such as a digital certificate, smart cards, and mobile phone. More recently, smartphones are being a potential alternative as a key enabler of secure authentication. Some of the latest smartphones include important security components such as a Trusted Platform Module that is able to secure digital certificates and cryptographic keys used for authentication. • *Something the principal is*, such as static biometrics (e.g. fingerprint, retina, and face) or dynamic biometrics (e.g. voice pattern, handwriting characteristics, and typing rhythm). On the other hand, the NIST 800-83 R4 SC-8-1 controls can be realised by AES or Triple DES; two approved symmetric algorithms.

collecting personal data, the system shall provide data subjects with the option to 'consent' and 'do not consent' the [processing instance].

Finally, while it should be recognised that the state of the art already provides plenty of controls contained in standard catalogues and frameworks for other more mature properties (mainly in the cybersecurity realm), controls related to anti-discrimination or autonomy are scarce and only recently there are some efforts and initiatives to address them (e.g. the EC has released ethics guidelines for trustworthy AI on April 8, 2019 [5]).

10.5 Recommendations

The European and international landscape of labels/seals is heterogeneous, as there is a great variation around their core functional models, the criteria they assess, the assurance level they offer, etc., and they also present a number of issues that need to be addressed [1]. For example, most of labelling core functional models require a complex chain of trust involving several third parties throughout the labelling process (e.g. evaluation body, certification/declaration authority, and accreditation authority). This complexity often results in a lot of time (and effort) required in the preparation and assessment of an ICT product/service, as well as in affordability issues due to the high costs involved. These issues are exacerbated when an ICT product or service must pass through the same process several times (one for obtaining the label and some other for certifying specific properties), involving additional cost and time. The industry has also highlighted these matters and called to "*minimize the burden on providers/manufacturers with respect to assessment, costs and time to market while ensuring an adequate level of trustworthiness*" [2].

While the TRUESSEC.eu labelling proposal advocates for addressing the complexity and affordability issues by reducing the intervention of third parties as far as possible, it also recognises the relevance of pursuing the verifiability and credibility of the labelling process. Providing the necessary evidence to support what is claimed about an ICT product or service improves verifiability. In turn, adding an independent public or private authority responsible for defining and articulating the labelling governance framework enhances credibility.

In this context, the TRUESSEC.eu proposal advocates a labelling solution that includes the following key elements: a self-assessment questionnaire, a labelling portal, a transparency report plus a visual label, and a governance framework ruled by an authority. Figure 10.5 illustrates the labelling approach proposed by TRUESSEC.eu.

- *The self-assessment questionnaire* is based on the indicators defined in the Criteria Catalogue. It provides a set of yes/no questions for a service provider to determine its compliance with the Criteria Catalogue. A

---

[5]European Commision, "Ethics Guidelines for trustworthy AI", https://ec.europa.eu/digital-single-market/en/news/ethics-guidelines-trustworthy-ai (accesed 12 April 2019)

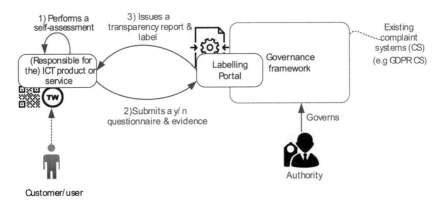

Figure 10.5   TRUESSEC.eu labelling proposal.

service provider performs the self-assessment and attaches the evidence of an indicator's fulfilment when the answer to a question is affirmative.

- *The labelling portal* processes the questionnaire answers and issues a transparency report and a visual label according to the level of conformance achieved for each of the twelve criteria included in the Criteria Catalogue.
- *The transparency report and the visual label* deliver a two-layer trustworthiness declaration. The visual label provides the first layer as it is easy to understand. The transparency report further details the assessment results in both text and machine-readable format, thus providing the second layer. The label and the report are both multi-dimensional (twelve criteria for trustworthiness) and multi-level (several levels of conformance for each criterion).
- *The governance framework* sets the fundamental rules the label must follow.

The following sections further elaborate on these elements.

10.5.1 Questionnaire

The questionnaire contains a set of yes/no questions, each asking whether an *indicator* of the Criteria Catalogue is met. The answer to each question allows to objectively determine which indicators are met and, ultimately, to what extent an ICT product or service meets a criterion for trustworthiness. Thus, we envisage a self-assessment and a yes/no questionnaire whereby providers/manufacturers reveal which *indicators* for trustworthiness they comply with, attaching the corresponding evidence when applicable. *Indicators* act as checkpoints, so they should be **empirically observable** (i.e. through evidence) and **objectively measurable** (i.e. a measurable element should be clearly defined in the indicators' description).

In our context, evidence refers to the information used to support the assessment and compliance of the *indicators*. Some evidences can refer to the implementation/realization of a given technique (e.g. the fifth indicator of the *'user-friendly consent' criterion: users are given the option to opt-out from data processing* can be supported by a centralized privacy control panel that includes opt-out options). Other evidences can describe organizational means to meet an indicator (e.g. those related to the *appropriate dispute resolution criterion*). Yet other evidence can be supported/provided by third parties who already performed an assessment on the subject-matter of the labelling e.g. through a certification or audit process. In this way, we prevent an ICT product or service from going through the same process several times (one for obtaining the label and some other for certifying specific properties). For example, a provider/manufacturer can link the certificate issued by a trusted third-entity as evidence of meeting the second indicator of the *'Cybersecurity' criterion: the ICT product or service is compliant with relevant [security] standards*.

An *indicator* should also include a measurable element easy to justify with evidence, calculate and understand. This measurable element should be clearly identified in the *indicator*'s description along with the corresponding measurement scale, which may be one of the following:

- **Nominal scales** are applicable for mapping values (without an intrinsic order) to categories, and only equality operation is allowed. The nominal **dichotomous** scale only has two categories and can be used to express whether a feature is present or not. In the Criteria Catalogue, several measurable elements are dichotomous in nature. For instance, the second *indicator* (ii) of the *'Cybersecurity' criterion* encloses a dichotomous measurable element with true or false as possible values. An evaluator will check whether the *ICT product or service is compliant with relevant [security] standards*. A provider/manufacturer can provide the certificate issued by a third trusted entity as evidence. Similarly, this can be applied for the first *indicator* of the *Privacy 'Commitment' criterion,* which states that "*The ICT provider clearly states its commitment to the GDPR in the form of a declaration*".
- **Ordinal scales** allow to sort or rank two or more categories, and equality and inequality operations are allowed. This may be applicable to, e.g.

the *'Enhanced control mechanisms' criterion.* The first *indicator* of this criterion states that *means to deletion of personal data should be provided.* In this respect, the *level of recovery* may be a measurable element intended to assess the difficulty (or easiness) to recover supposedly deleted data. For example, based on the guidelines and techniques presented into the NIST SP 800-88, three values on the ordinal scale can be abstracted:

- Level 1 (Clearing) – Deletion is done using overwriting software not only on the logical storage location but on also all addressable locations, so data cannot be easily recovered with basic utilities but could be possible with laboratory attacks.
- Level 2 (Purging) - Deletion is done using sophisticated sanitization techniques, so data cannot be possible at all.
- Level 3 (Destroying) – The media is destroyed (physical destruction).

Therefore, this measurable element can have three different ordinal levels, and the assurance of a given *level of recovery* can be an *indicator* attached to a particular *level of conformance.* For example, ensuring the Level 1 (Clearing) can be a criterion of the Level of Compliance 1, and the corresponding successive levels.

- **Interval/ratio scales** have numerical values and allow obtaining the difference or distance between them allowing be comparing and ordering. This may be applicable to measurable elements that have continuous numerical values. For example, the *period for the disposal of personal data once they have been processed for the purpose consented to* be another relevant, measurable element of the criterion mentioned in the previous paragraph. This period may have a continuous and infinite range of values, e.g. 1 day, 30 days, 365 days, etc. These quantitative, measurable elements can then be embedded in dichotomous (yes/no) *indicators* in terms of intervals or thresholds. As a matter of example, an *indicator* belonging to an advanced *level of conformance* may state that personal data are automatically deleted as soon as they are not used (0 days), while an *indicator* of *a basic/entry level of conformance* may state that personal data are deleted within 15 to 30 days.

10.5.2 Labelling Portal

Based on the answers submitted by a provider/manufacturer, the labelling portal issues a transparency report and a visual label conveying the *level of trustworthiness* of the ICT product or service assessed. The notion of *level of trustworthiness* must be understood neither an absolute "yes/no trustworthy" nor as a single scalar "75.5% trustworthy", but as the extent to which the twelve criteria for trustworthiness defined in the Criteria Catalogue are fulfilled. This "extent" corresponds to one of the *levels of conformance* defined in the labelling scheme. To illustrate the notion of *level of conformance* and supported by the levelling structure defined in and [9], the following levels have been defined: Basic/Entry (Level I), Enhanced (Level II), and Advanced (Level III).

We advocate for an assessment based on groups of *indicators*, where each group is associated with a qualitative *level of conformance*. As illustrated in Figure 10.6, the *indicators* of each criterion are divided into subsets. Each subset is assigned to a particular Level of Conformance. For an ICT product or service to reach a superior *level of conformance* in any of its criteria, it must necessarily comply with all the *indicators* of the previous levels. Therefore, a criterion that has a Level I means that it complies with all the *indicators* belonging to Level I, Level II implies that a criterion complies with both Level I and Level II *indicators*, and Level III implies that a criterion complies with Level I, Level II, and Level III *indicators*.

An ICT product or service can have different *levels of conformance* for each of the twelve Criteria for trustworthiness. Figure 10.6(b) further depicts two items (ICT products or services) in its last two columns. On the one hand, the item A complies with Level II for criterion C1 (Information) and with Level I for criterion C2 (User-friendly consent). On the other hand, item B complies with level I for criterion C1 and level III for criterion C2. Also, note that item B does not conform to level II in criterion C1 because it fails to meet *indicator* I1.4. In this example, it can also be noted that if an ICT product or service is not able to comply with a single *indicator* for some level, it does not conform to that level.

The decision to define different levels for the Criteria is supported by different legislation; for example, the European GDPR (General Data Protection Regulation) defines different degrees of sensitivity of personal information, each requiring different privacy controls to protect them. Therefore, different privacy protection controls could be mapped to different subsets of *indicators*, each assigned to a respective *level of conformance*. Similarly, the Cyber Security Certification Framework by European Commission defines three Assurance Levels, each assigned to different subsets of requirements/criteria in terms of the risks involved.

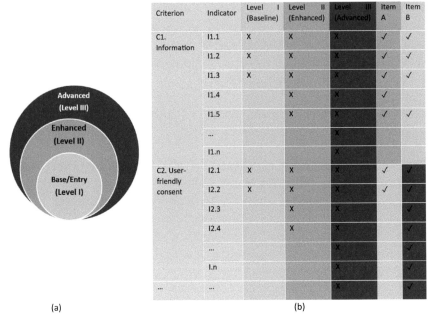

Criterion	Indicator	Level I (Baseline)	Level II (Enhanced)	Level III (Advanced)	Item A	Item B
C1. Information	I1.1	X	X	X	✓	✓
	I1.2	X	X	X	✓	✓
	I1.3	X	X	X	✓	✓
	I1.4		X	X	✓	
	I1.5		X	X	✓	✓
	...			X		
	I1.n			X		
C2. User-friendly consent	I2.1	X	X	X	✓	✓
	I2.2	X	X	X	✓	✓
	I2.3		X	X		✓
	I2.4		X	X		✓
	...			X		✓
	I.n			X		✓
	...			X		✓

(a)   (b)

Figure 10.6 (Illustrative) Levels of conformance.

10.5.3 Transparency Report and Visual Label

The trustworthiness of an ICT product or service is expressed as twelve dimensions (criteria) each with a level of conformance depending on the subset of indicators met. These are conveyed to the label consumer through a two-layer declaration:

- The first layer shows a *visual label* that is easy for users to understand. It shows the extent to which each trustworthiness criteria is fulfilled (i.e. criterion plus its level of conformance).
- The second layer shows a *transparency report* in both text and machine-readable format. This should provide further details, i.e. criteria, *indicators* fulfilled, evidence provided (if applicable), and the individual *levels of conformance*. The machine-readable *transparency report* enables machine-to-machine integration based on e.g. the users' policy settings as set in their user agents such as a web browser. This may facilitate the automation of products and services trustworthiness comparison and assessment.

Both the transparency report and the visual label should highlight the date of the last update and should clearly specify which components of the product (modules/functionalities) or service (operations) are part of the labelling. In addition, in order to verify the authenticity of a label the following measures need to be considered:

- The *labelling portal* (who issues the *transparency report* and the *visual label*) should publicly provide a list of issued labels, including the two-layer information above described.
- The *visual label* also should integrate a link to forward the user to the *labelling portal*, which provides information about the corresponding ICT product and service.
- The authenticity of the *labelling portal* should also be ensured.
- The Criteria Catalogue should be easily accessible to the public, i.e. freely downloadable from a public website.

10.5.4 Governance and Authority

Having an independent third party managing the verification of criteria/indicators and subsequent declaration increases the credibility and ultimately the degree of user confidence in a labelling scheme, since, e.g., fraudulent behaviour or user complaints are managed by these independent entities. This is supported by previous findings [1] which suggest that (i) the schemes operated by public bodies or foundations were found to be the most transparent, comprehensive and, trustworthy; and, (ii) labelling schemes have poor longevity unless they are backed by public authorities or large operators. Thus, the TRUESSEC.eu labelling solution advocates for a governance framework ruled by a public or private authority that will be responsible for:

- Creating the yes/no questionnaire.
- Deciding the number of *levels of conformance.*
- Assigning indicators to each level of conformance.
- Setting a validity period for the *transparency report* and the *visual label*. It should be considered that the Cybersecurity Act states that certificates shall be issued for a maximum period of three years and may be renewed, under the same conditions (Article 48). The same is stated by the GDPR (Article 42). However, the 'lightweight nature' of the proposed labelling solution allows re-issuing the *transparency report* and *visual label* in shorter time thus increasing the credibility of the approach. Therefore, we recommend a 12-month expiry date from the last update.

- Defining the terms and conditions on the use of a label. This should include penalty rules in case of cheating or non-compliance as well as supervision mechanisms to ensure the validity of the label (e.g. random audits or complaint channels). In this sense and aligned with our "re-use and no-burden approach":

  - We recommend that penalty and complaint approaches already defined in other close legislation, e.g. GDPR, are considered and articulated with the labelling system here proposed. Some 'Core Areas of Trustworthiness' fall within already regulated areas (e.g. privacy and security). Therefore, considering, e.g. that most of the *indicators* in the Criteria Catalogue are covered by the GDPR, its complaint and penalty regime (GDPR CHAPTER VIII: Remedies, liability and penalties) should be articulated with the labelling system. Thus, e.g. a GDPR breach will trigger a re-issue of the *transparency report* and *visual label* (in this case, even the basic/entry level would not be met).

  - Non-compliance with a criterion should not necessarily result in the revocation of the label, but its update to reflect a new *level of conformance*. Revocation should only be performed when at least the basic/entry level is not met.

10.6 Conclusions

The current world scenario shows that the users feel unable to recognise the level of trustworthiness of applications and services, and not even identify which characteristics should they have or show, depending on the confidentiality or sensitivity of the process the user is intending to perform with them.

This makes users feel helpless facing the dilemma "to trust or not to trust".

In this scenario, the trust labels appear to be the solution, i.e. the users could look at the label issuer, and ask its experts to take a decision on their behalf, or at least make some assessment of the level of trust on the application the user could make, in one or several of the criteria identified in TRUESSEC.eu.

This scenario is a somewhat utopic for several reasons:

- There are not well recognised trustworthiness labels, so the users don't know about its existence
- Which ones they should trust more, based on the specific user requirements and expectations about the behaviour of a specific application.

- Which levels of trust and on which areas should the user request from the application or service provider.
- Who evaluates the level of trust of the applications and on which criteria, to assess the level of trust, so that the users could be confident that the assessment itself is trustworthy.

In order to change this pessimistic scenario, the first thoughts of the project in order to propose a roadmap for the implementation of a trustworthy widely adopted trust label (or set of), are taking into consideration the following ideas:

1. Involvement of well-known and authoritative stakeholders, like ENISA, FRA or other European Union institutions, issuing and supporting recommendations to launch and promote the adoption of the trustworthiness label(s).
2. Encourage organisations active in the cybersecurity awareness, like APWG.eu and most of the EU Member States N/G CERTs, to disseminate and make the citizens aware of the existence and advantages of using those trustworthy labels for their own cyber-safety.
3. Define a methodology to allow application developers and service providers to self-assess the trustworthiness of their applications in some or all the criteria identified in TRUESSEC.eu. This approach is aligned with the policy adopted by ENISA in the PET assessment tool. Adoption of this strategy by application developers and service providers will be proportional to the effective demand expressed by the users in the Market.
4. National and/or European authorities should appoint a supervisory authority that could validate the accuracy of the self-assessment statements made by developers and service providers, in order to provide the required trustworthiness to the whole assessment schema. Optionally the assessment criteria could be upgraded to standard and be evaluated by an independent laboratory or trusted third party, which would provide an additional level of trust on the label by the citizens.

Acknowledgements

The research leading to these results has received funding from the European Union's Horizon 2020 research and innovation programme under grant agreement No. 731711. The first author gratefully acknowledges his sponsor, Escuela Politécnica Nacional.

References

[1] V. Gibello, "TRUESSEC Deliverable D4.1: Legal Analysis," 2017. [Online]. Available: https://truessec.eu/content/deliverable-41-legal-analysis.

[2] V. Gibello, "TRUESSEC Deliverable D7.1: Evaluation of existing trustworthiness seals and labels," 2018. [Online]. Available: https://truessec.eu/content/deliverable-71-evaluation-exiting-trustworthiness-seals-and-labels.

[3] H. Stelzer, E. Staudegger, H. Veljanova, V. Beimrohr, and A. Haselbacher, "TRUESSEC Deliverable D4.3: First draft Criteria Catalogue and regulatory recommendations," 2018. [Online]. Available: https://truessec.eu/content/d43-first-draft-criteria-catalogue-and-regulatory- recommendations.

[4] D. S. Guamán, J. M. Del Alamo, H. Veljanova, S. Reichmann, and A. Haselbacher, "Value-based Core Areas of Trustworthiness in Online Services," in IFIP International Conference on Trust Management, 2019, Springer, Cham.

[5] D. S. Guamán, J. M. Del Alamo, H. Veljanova, A. Haselbacher, and J. C. Caiza, "Ranking Online Services by the Core Areas of Trustworthiness", RISTI-Revista Ibérica de Sistemas e Tecnologias de Informação, 2019.

[6] L. Floridi, "Soft Ethics: Its Application to the General Data Protection Regulation and Its Dual Advantage," Philos. Technol., vol. 31, no. 2, pp. 163–167, 2018.

[7] D. S. Guamán and J. M. del Alamo, "TRUESSEC Deliverable D5.2: Technical gap analysis." [Online]. Available: https://truessec.eu/content/deliverable-52-technical-gap-analysis.

[8] D. S. Guamán, J. Del Álamo, S. Martin, and J. C. Yelmo, "TRUESSEC Deliverable D5.1: Technology situation analysis: Current practices and solutions," 2017. [Online]. Available: https://truessec.eu/content/deliverable-51-technology-situation-analysis- current-practices-and-solutions.

[9] European Cyber Security Organisation, "European Cyber Security Certification: A meta-scheme approach," WG1 – Standardisation, certification, labelling and supply chain management, 2017. [Online]. Available: http://www.ecs-org.eu/documents/uploads/european-cyber-security-certification-a-meta-scheme-approach.pdf. [Accessed: 05 April 2018].

11

An Overview on ARIES: Reliable European Identity Ecosystem

Jorge Bernal Bernabe[1], Rafael Torres[1],
David Martin[2], Alberto Crespo[3], Antonio Skarmeta[1], Dave Fortune[4],
Juliet Lodge[4], Tiago Oliveira[5], Marlos Silva[5], Stuart Martin[6],
Julian Valero[1] and Ignacio Alamillo[1]

[1]University of Murcia, Murcia, Spain
[2]GEMALTO, Czech Republic
[3]Atos Research and Innovation, Atos, Calle Albarracin 25, Madrid, Spain
[4]Saher Ltd., United Kingdom
[5]SONAE, Portugal
[6]Office of the Police and Crime Commissioner for West Yorkshire, (POOC), West Yorkshire, United Kingdom
E-mail: jorgebernal@um.es; rtorres@um.es; martin.david@gemalkto.com; alberto.crespo@atos.net; skarmeta@um.es; dave@saher-uk.com; juliet@saher-uk.com; tioliveira@sonae.pt; mhsilva@sonae.pt; stuart.martin@westyorkshire.pnn.police.uk; julivale@um.es; ignacio.alamillod@um.es

Identity-theft, fraud and other related cyber-crimes are continually evolving, causing important damages and problems for European citizens in both virtual and physical places. To meet this challenge, ARIES has devised and implemented a reliable identity management framework endowed with new processes, biometric features, services and security modules that strengthen the usage of secure identity credentials, thereby ensuring a privacy-respecting identity management solution for both physical and online processes. The framework is intended to reduce levels of identity-related crimes by tackling emerging patterns in identity-fraud, from a legal, ethical, socio-economic, technological and organization perspective. This chapter summarizes the main goals, approach taken, achievements and main research challenges in H2020 ARIES project.

11.1 Introduction

In a world getting every time more and more digital, the protection of the personal data is a crucial point, in particular, individual identities are vulnerable in this scenario, where European stakeholders are interacting in a global way. The lack of trust is increasing derived from the current absence and deficiency of solutions, including consistently applied identification and authentication processes for trusted enrolments, particularly the use of online credentials with low levels of authentication assurance. Moreover, there is not a common approach in Europe (from the point of view of the legislation, cross-border cooperation and policies) to address identity-related crimes. This situation costs billions of Euros to countries and citizens in fraud and theft.

In this scenario, ReliAbleeuRopean Identity EcoSystem (ARIES) H2020 research project aims to provide a stronger, more trusted, user-friendly and efficient authentication process while maintaining a full respect to subject's and personal data protection and privacy.

Thanks to this ecosystem, citizens will be able to generate a digital identity linked to the physical one (eID/ePassport) using biometrics while, at the same time, store enrolment information in a secure vault only accessible by law enforcement authorities in case of cybersecurity incidents. Because of this process, linking proofs of identity based on the combination of biometric traits and citizen digital identity with the administrative processes involved in the issuance of documents like, for example, birth or civil certificates will be possible.

Users will also be able to derive additional digital identities from the ones linked with their eID or ePassports with different levels of assurance and degrees of privacy about their attributes. The new derived digital identities may be used in administrative exchanges where it is required by the governments according to eIDAS regulation [1] and be store in software or hardware secure environments in their mobiles or smart devices.

The rest of this chapter is structured as follows. Section 2, depicts the ARIES Ecosystems, Section 3 is devoted to the main innovative process in ARIES. Section 4 describes the legal and ethical approach considered. Section 5 recaps the main cyber-security and privacy Research challenges. Finally, Section 6 concludes this chapter.

11.2 The Aries Ecosystem

The project goal is to provide new technologies, processes and security features that ensure a higher level of quality in security aspects like credential management for privacy-respecting solutions and the reduction of identity fraud, theft or wrong identity problems which can be associated with crimes. The general Aries ecosystem is depicted in Figure 11.1.

Authentication processes will be ensured with the use of smart devices allowing to use all required biometric (especially face) and electronic (using NFC) data. This process should ensure a high level of quality for biometrics acquisition, while assuring data integrity and delivering the derived identities required attributes to the adequate relying party (service provider). Such features will be achieved by functionality locally (on the smart device) or centrally (back-end). Moreover, digital identities will be generated with privacy preserving technologies and allowing citizens to just prove to be in possession of some attributes without exposing the rest of their data, i.e. being over 18 years old. Given that different levels of assurance are possible a biometric mechanism could also be used as a proof of digital identity possession where appropriate.

A user manages multiple identities and credentials which are issued by Identity Providers (IdP) and presented to the Service Providers (SP) to access the offered services by them. The ARIES approach considers a multi-domain interaction for eID management in order to achieve a distributed but unified eID ecosystem. Each domain usually contains one or more IdPs and one or

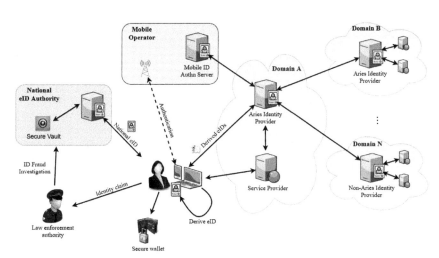

Figure 11.1   ARIES ecosystem overview.

more SPs. Usually, a SP redirects the user requests to the IdP within its own domain but there are some exceptions to be considered: An SP can directly authenticate the identity of the user (e.g. validating a certificate) and could, also, redirect to an IdP of another domain in which it trusts, including a mobile operator, a bank or a Government for Mobile eID authentication. In addition, IdPs can be interconnected relying on federated interoperability, thereby allowing delegation of authentication (e.g. using STORK) and also attribute aggregation (e.g. to create a derived credential which includes both governmental and academic information). User consent will be obtained prior to transferring any personal information. Interaction with legacy non-ARIES IdPs can be also achieved by contacting those IdPs via standard protocols such as SAML [2], OAuth2 [3], etc.

The users interact with the system through several devices such as mobile phones or smart wearables. These devices will require a secure element in order to securely protect digital identities with biometric features. Alternative, although less secure, storage and execution environments might be foreseen for larger adoption of the ecosystem, but with limited capabilities to manage the resulting risk. A secure electronic wallet will be provided to users for them to securely handle and manage their digital identities and their related data.

New derived credentials can be requested by users to an ARIES IdP after an authentication process using its eID. These credentials may contain different identity attributes and/or pseudonyms, according to the user needs and required level of privacy and security. The issuance process implies that the new credential must be issued by a trusted IdP and logged to assure the traceability in law enforcement purposes such as real identity identification. This could be achieved by an encrypted and signed logging mechanism. The logged information should also be kept secured and only accessed due to law regulated cases.

The derived credentials will be originated from previous strong ones such as biometric data and eID documents. The process is based on mobile token enrolment server with a derivation module. Optionally, users could also derive their own identity and present cryptographic proofs to an SP. In this case, identity approaches based on Zero Proof Knowledge Proofs like IBM Idemix [4] or ABC4Trust [5] solutions can be used. For this, the obtained credentials should be prepared with the needed cryptographic information to derive new identities and provide proofs when requested by an SP.

Those identities can be derived and issued by different entities having, each credential, an associated Level of Assurance (LoA). This serves as a measure of the security mechanism used by the credential issuer to validate the user identity. ARIES aims to keep the LoA or to avoid significant differences when using derived credentials and similarly, will try to ensure that Level of Trust (LoT) among different entities is also a maintained after adopting derived credentials in the ecosystem.

Accessing a service provided by an SP will impose some requirements for the credential to be presented, like including some attributes about user identity and other trust requirements. The user can choose the mechanism and the credential which he wants to present according to his preferences and the required information by the SP. This includes the usage of derived credentials or pseudonyms with less exposed information; a proof of identity in which no credential is actually sent to the SP, but a proof that the user owns some identity or attribute; or a Mobile ID credential stored in a secure element, which makes use of the Trusted Execution Environment for authentication and can optionally involve mobile operator as party involved in circle of trust [6].

Privacy by design principles is essential from the data protection perspective especially when identifying which and how biometric data are going to be used. Indeed, the requirements of proportionality have to be analysed, bearing in mind the demands of technical security measures, determining what is certainly essential to avoid identity thefts based on the access to biometric information (e.g. a photo in the case of face or a latent in the case of fingerprints).

Likewise, the possibility of using several derived identity credentials demands a concrete assessment from the perspective of data protection. Therefore, it will be necessary to build up identification services prioritizing those technical and organizational solutions that minimize access to personal data to the absolute essential. To accomplish these objectives, ARIES devises means to comply with the minimal disclosure of information principle. In this sense, the principle of proportionality will play a key role in order to face this challenge, since it will be necessary to justify in each case by the service providers the personal data really required for authentication or authorization.

Furthermore, the identity ecosystem will provide unlinkability at the relying party level through polymorphic user identifiers (when compatible with relying parties' authentication policies). For each authentication or for specified periods of time will be different and random identifiers, so it will disclose no information. The unlinkability at the ARIES IdP will be also ensured. The ecosystem will indeed hide the accessed service from the enrolment and authentication services. Unobservability will be ensured by the system architecture as well. The Identity Providers will have no information about which SP the user wants to log into.

11.3 Main Innovative Processes in Aries

11.3.1 Fraud Prevention and Cyber-crime Investigation

The topics of fraud and crime prevention and investigation were one of the main project goals and were addressed from the very beginning of the project by involving law enforcement personnel in the process of requirement definition. Their inputs were based on currently most frequently occurring threats and their experience from crime investigation, and based on them an assessment of state of the art authentication architectures was done at the beginning of the design phase.

Main results of the assessment resulted in three main improvements in the field of crime prevention and investigation the project may provide:

– Strong authentication accessible to large part of the population to replace legacy authentication types (such as password or SMS one-time code).
– Biometric authentication as additional obstacle for the criminals.
– ID Proofing with document reading and biometric verification as a strong identity verification means to ensure the newly issued privacy preserving partial identities are based on reliable information.

It is obvious the strength of the whole solution depends on algorithms used for biometric verifications (both live capture vs. image data from electronic document and live capture vs. previously enrolled baseline template). This was in line with the project plan as improvement of both enrolment and verification of face biometrics was planned as a separate task.

If the authentication is broken by some means and the investigation takes place it usually requires as much information as can be obtained (IP address, device fingerprints, all transaction data) which is in contradiction with project goal to provide privacy friendly solution. It was decided the privacy and a control over user's own data is more important than inputs for investigation and a limitation was introduced. User data stored inside of the system are encrypted and stored in an appliance called Secure Vault. The appliance enforces strong authentication and authorization based on ARIES authentication, so it is only the user himself who can approve access to his data. This limits the investigation to cases when user's identity was stolen by forgery, but he still has access to his mobile phone to be able to provide access to law enforcement authorities.

The data collected and stored by all server-side components of ARIES solution consist of transaction information and anonymized identity information such as links between the cryptographic and biometric identity parts.

It was decided to introduce a rule that biometric information may be persisted only in user's handset in order to give him control over this most important information.

11.3.2 Biometric Enrolment and Authentication

The architecture considered that the biometric authentication is evolving and new methods and implementation are introduced as fast as the old ones are broken by new approaches such as deep learning. The solution is a set of loosely coupled server components that allows simple replacement of each component without much impact on the existing ones. To integrate with ARIES each vendor must provide server side API and App SDK, the project provides enveloping App with UI flow control and a server application that controls the issuance and authentication flows and ensures all steps happen in a single session. The choice of which biometric feature should be enrolled is based on user's choice and his handset capabilities.

The implemented OpenID Connect authentication flow allows selection of any available biometric authentication type, so the requesting Service provider may choose the optimal balance between level of assurance and user experience that may be worse for some biometric features.

The project considered two options of biometric authentication: usage of the feature obtained during ID Proofing process (at the moment only face image is accessible for commercial applications) and usage of another feature as an additional authentication factor without link to the original electronic document information.

The face recognition done during ID Proofing strongly relies on quality of the image from electronic document. During project pilot phase issues with several passports with poor quality image data were encountered that prevented enrolment of the users. The liveness detection is in ID Proofing mandatory, because if the attacker has stolen document then he gets hold of the image data himself, so the liveness detection is the only protection.

Pilot implementation used face recognition combined with ID Proofing and the results were satisfactory:

– Enrolment was successful for majority of the users and was done without issues on the first try.
– Authentication with liveness detection based on head movements (vertical and horizontal) using overlay image to tell the user what to do was smooth and well accepted by the pilot users. Average face verification time was below 3 seconds.

Voice authentication was implemented to prove the solution is able to quickly integrate an existing biometric authentication service. Existing server-side service proposed by one of the partners was selected and integrated in two steps: scaffolding REST service was created to align the API style and session management and very simple App SDK was implemented and added into existing ARIES App.

11.3.3 Privacy-by-Design Features (Anonymous Credential Systems)

ARIES follows a privacy-by-design approach to protect user's privacy in their digital transactions, either online or offline (on-situ, face-to-face interactions). The architecture has been designed to incorporate and interface with Anonymous Credential Systems (ACS), namely Idemix [4]. ACS allows users to set-up and demonstrate Zero Knowledge crypto Proofs, thereby proving certain predicates about personal attributes in a privacy-preserving way, following a selective disclosure approach.

The ARIES Mobile App allows obtaining ACS credentials, once the user has been identified and enrolled in the ARIES IdP. Those credentials are generated based on the attributes demonstrated by the user against the IdP during the ID-proofing, i.e., it contains at least the attributes included in the breeder document (ePassport) used for authentication and enrolment. The credentials are maintained securely protected inside the mobile (mobile wallet).

Once the user has performed the issuance protocol, it can create different proofs of possessions to comply with attributes required by the Service Provider to access a service. This presentation protocol is based on ZKP by relying on the CL signature scheme [7]. It ensures minimal disclosure principle, allows demonstrating having an attribute without disclosing the value itself, and permits proving complex predicates about attributes, e.g. the date of birth is greater than certain year (to check age). Anonymous credentials systems have been also integrated, and successfully evaluated, for IoT scenarios [8] in even constrained IoT scenarios [9] in the scope of Aries project.

In ARIES these privacy-preserving capabilities have been showcased in the Airport scenario, in which the user wants to demonstrate he is over 18 to buy certain products (e.g. alcohol) in a duty-free shop inside the airport, and prove that he has a valid boarding pass (required to buy goods) without revealing any personal data, proving only he is traveling to a valid destination in a valid time-frame.

11.4 The ARIES Ethical and Legal Approach

11.4.1 Ethical Impact Assessment

ARIES focussed on how to optimise the potential for minimising and averting unintended misappropriation and disproportionate use of information for unknown and diverse purposes to which citizens have not explicitly consented. It did so in ways that bring privacy and security in balance while addressing the socio-ethical consequences of deploying the ARIES solution to creating a reliable, trustworthy eID ecosystem.

ARIES also has as its foundations the EU's ambitious commitment to realising a Single Digital Market relies on creating trustworthy eIDs to augment efficiency, convenience and trustworthiness of e-life for citizens. Digital by default is enabled by the once-only principle (to cut multiple entry of same data several times), interoperability by default, inclusive and accessible practices.

11.4.2 Technological Innovation Informed by Ethical Awareness

Technological innovation is not neutral in conception, in development or in its application to society. Algorithms are not neutral. This is the starting point for reflecting on the ethical tests that might be applied as a new application is developed or an existing one extended and used for a different, but possibly complementary purpose, to the one for which it was first developed. Just because a development or application meets current legal privacy requirements, it cannot be assumed that it automatically complies with ethical standards that society values. This means that there must be clarity over the purpose of a new development and its intended used in real-time and in the real-world. The legal tests of ensuring compliance with the law provided by privacy impact assessments themselves are useful check points. By themselves, they are inadequate. Legal compliance is necessary but not sufficient to ensure ethical standards are met.

11.4.3 The Socio-ethical Challenge

The key challenge for ARIES was to develop something that was universally acceptable, complied with legal and ethical requirements, while protecting security and privacy and would help form the basis of a reliable and trustable eco-system. Accordingly, ARIES set about developing a neutral application that sought to facilitate convenient, privacy respecting, secure, and speedy transactions whilst minimising the amount of personal data that an individual citizen might be required to disclose (by choice or design) in order to access a service.

11.4.4 ARIES Starting Point: What is Meant by Ethics?

The ARIES ecosystem is designed with both privacy and ethics in mind. ARIES extracted core principles of ethical practice from philosophy and medicine which have addressed the impact of technical and scientific advance on what it is to exist as a human being. There is no universal acceptance of what is ethically appropriate or acceptable. Consequently, designing something that is 'ethical by design' implies designing something that minimizes objections to it from different societies and is an essential building-block of an ethical e-ID eco-system.

The do no harm principle provides the best initial ethical test to be applied to the design of a new algorithm or app. It is useful for a digital society accustomed to automated decisions being driven by bots rather than immediate live human decision making on a human2human basis. However, this immediately raises additional ethical issues summarized by the principles of proportionality, purpose specification and data and purpose minimization. Dignity and autonomy are core elements of the concept of bodily integrity. To those are added notions of privacy (in private and public transactions).

In short, the use made of something, like an eID, occasions many ancillary questions about the person associated with it. This is problematic and has preoccupied legislators and citizens anxious to ensure that they do not inadvertently reveal and allow to be sold for commercial gain, aspects of themselves (i.e. data and associated information that they generate). Further ethical issues arise. Therefore, ARIES seeks to develop a solution which bakes in ethics and is as neutral as possible in its impact on societal values.

11.4.5 Embedding the Dominant Ethical Principle: Do No Harm

The key ethical principle to which all other ethical principles are linked and subordinate is the pre-cautionary principle. It highlights the obligation to 'do no harm'. Closely associated and derived from it are principles impelling proportionality, self-determination and consent, autonomy, dignity, and necessity (data minimisation). Refining accepted medical ethics for informational technology practices suggests that ethical practice and ethical technological applications need to be aware of, and in the case of eIDs, sensitive to how they will mitigate, avert or accommodate risks (or potential harms).

The precautionary principle of do no harm is about more than determining legal liability and redress for harm. In ARIES, it informed design and practice from the start. This differs from the traditional practice of using legal remedies for harms, and the focus in the USA, for example, of litigation to provide financial recompense for harm. In ARIES, attempts were made to widen the understanding of what 'harms' might be induced by ICT innovations in line with the EU approach to baking in the precautionary principle of 'do no harm'. In the EU, this is expressed in guidelines and in legislation which translates this principle into duty-of-care provisions, as in the case of the GDPR and the complementary ePrivacy Directive (soon to be Regulation). This duty-of-care has been marked in respect of privacy protection in both the GDPR and ePrivacy deliberations: both require importers and retailers of IT to distribute only privacy-by-design compliant technology. The temptation to assume that PbD compliance automatically implies respect for ethical principles must be avoided.

For ARIES ecosystem, ethics is seen in relation to when, how and by whom (or what algorithm) decisions are made, and for what purpose. This means that there are several points at which ethical reflection must occur in order to guard against baked in bias and ensure ethical principles are respected in terms of all elements of the design, from inception to roll-out, to scalable use. Such ethical checks occur at the following points: design of the medium in which personal information is to be held; technical rules governing handling or and access to that information, including via a human or bot; technical vulnerability to the integrity of the medium and its message; commercial opportunities; and impact on the individual providing information (knowingly or not) to access a service. ARIES reflected on how ethical principles may be used to inform data handling practices that rely, at some point, on eID authentication on the part of the individual or the service provider.

Core ethical principles to be observed are: precaution; proportionality; purpose specification; purpose limitation; privacy; security; autonomy; dignity; informed consent; justifiability; fairness; transparency and equality.

11.4.6 Baked in Ethics for the ARIES Use Cases

The ARIES Use Cases on eCommerce and e-Airport reveal that different rules apply to eID based transactions in a common physical setting owing to pragmatic and political constraints imposed by real-world contexts, real-time eID development and use. These values, so far, are shaped by human beings.

For the ARIES, the baseline was the ethical principles common to our societies in the EU28, awareness of what the public interest is; how it can be explained and protected. This entailed learning from on privacy assessment initiatives, regulations, oversight mechanisms, audit, inspection and compliance arrangements and independent scrutiny to ensure accountability and redress. Implicit are ethical principles of good governance, transparency of intent and effect. This places a premium on minimum disclosure requirements in terms of how algorithms are designed and used, phased and shaped (often by other automated processes) and deployment that is proportionate to the goal they are designed to attain. Ethical compliance is not met, therefore, simply by assuming, especially in the case of eCommerce, that competition and anti-trust legislation, standards and regulations are sufficient to guarantee ethical use. Nor is accountability just about liability for malfunction or misuse. This is why the baking in of ethics, an ethic audit trail even, mean that accountability has to be citizen focused and relate also to the intended use and effect of using an eID on society.

Ethical eID design therefore must reflect principles of accessibility, dignity, equality and transparency. Ethical design suggests that in practice where eID use fails to be used, for whatever reason, there should be clarity over why this happens and, in order to preserve dignity and accessibility, alternative means of completing an intended benign transaction. The GDPR Art 22 states that people have a right NOT to be subject to a decision 'based solely on automated processing'.

11.4.7 Ethics in the ARIES Use Cases

ARIES Use Cases rested on the same set of questions and methodological approach to ensure consistent application across all of the ARIES activities. All checked fitness-for-purpose. How is ethical use designed into the system? What bias is there? How can risks and benefits be reconciled? How have ethics been designed into the technical solution envisaged? Is this sufficient from the point of view of user trust building? ARIES was especially mindful of the inherent risk of doing inadvertent harm. Its Ethical Impact Assessment tool therefore reflects this by highlighting that any data enrolment, collection or (manual or automated) processing must not harm the data subject directly or indirectly. It must be proportional to the purpose for which processing occurs, must minimise data used and ensure that it is used for that one, specific and limited purpose only. No more data should be enrolled or collected and associated than is expressly necessary for the transaction envisaged.

Breaching the spirit of privacy preservation under the GDPR is a breach of ethical practice.

ARIES concludes that an EIA is a commercial opportunity in its own right and key to building sustainable trust and reliability while maximising privacy and security. An EIA should be conducted in parallel with PIAs. An Ethics audit via an independent and expert body should complement a PIA. These should be done before the decision to proceed with further development of the technical solution is taken. It must be done at the outset (possibly after taking external, independent advice) and authorised and signed off at the highest level. This helps create a trusted privacy and ethics respecting environment for developing innovative technical solutions. Ethically informed good practice becomes second nature. This is communicated to the public and stakeholders. Public trust is key to sustaining trust in the reliability, security and dependability of the solution.

11.4.8 Legal Challenges and Lessons Learned in ARIES

As explained in precedent sections, ARIES proposed the use of new identification techniques, fully user centric, that required a complete review of the different legal framework that may be considered applicable to the service, in case of real exploitation.

First of all, an analysis of the eID European Union legislation, and its application to the ARIES ecosystem was conducted, mainly focused in the Regulation (EU) N° 910/2014 of the European Parliament and of the Council of 23 July 2014 on electronic identification and trust services for electronic transactions in the internal market and repealing Directive 1999/93/EC (commonly known as "eIDAS Regulation") and its implementing acts.

Our findings include that an ARIES provider may play two different roles in the eID EU regulated ecosystem:

– First of all, an ARIES provider may be an electronic identification means consumer. This happens when the ARIES provider uses the electronic identification means issued to the citizen i.e. by the Member State, such when the citizen authenticates using a national citizen ID card (i.e. the Spanish National ID card, or the German nPA). This is an interesting way to reuse strong authentication-based identification mechanisms as an authentic source for the self-issuance of user-controlled identities.

– Secondly, an ARIES provider may be an electronic identification means issuer, in the sense of the eIDAS Regulation. For this to happen, the system must comply with the legal requirements set forth by the eIDAS

Regulation, and the corresponding implementing acts, and be recognized by a Member State. The ARIES derived identities aim to be recognized according to the substantial security level defined in Article 8 of the eIDAS Regulation and, thus, the system shall comply with the corresponding requirements set forth in the eIDAS Security Regulation. This possibility would allow the usage of the ARIES derived identities for the electronic access to public services in the EU.

Due to their nature as a private, pseudonymous, identification means with legal value under eIDAS Regulation, an analysis of the use of advanced electronic signatures based in qualified certificates issued by ARIES providers was also considered relevant, for the endorsement of derived identities. Research concluded that an ARIES provider may issue qualified certificates assuring the identity of the person, using pseudonym certificates and other attributes, as a means to represent derived identities. This possibility is directly implementable in the current EU framework, but its recognition is subject to the authorisation of the usage of pseudonym certificates in each Member State.

More interestingly, our research showed that an ARIES provider could offer a new trust service, consisting on the accreditation of possession of personal attributes (a wide conceptualization of identity) with privacy protection.

This may be considered as the main legal innovation of the project: an ARIES provider, once a person identity has been provisioned, provides a service that allow that person to self-create partial, derived, identities asserting in a trustworthy manner a particular personal attribute (i.e. the possession of a personal, valid, boarding pass to shop in the airport, or being older that certain age...). These derived identities constitute assertions that may legally substitute the corresponding documents that evidence the personal attributes (i.e. instead of showing the boarding pass, with all personal data, one shows a partial, derived identity that proves the fact that the person has a personal and valid boarding pass), thus increasing privacy effectively, while reducing compliance costs to data controllers.

To be able to substitute these documents per partial derived ARIES identities, maintaining legal certainty, a definition of this services a new trust service should be proposed, including the institutionalization of the service and a legal effect attained to the service (i.e. establishing some sort of equivalence principle such as "where the law requires the documental accreditation of a personal attribute, it will be possible to use a [service name] evidence".

11.5 ARIES Ecosystem Validation

11.5.1 E-Commerce

The secure eCommerce scenario focused on demonstrating how virtual identities with different levels of assurance can be used to access different online services. It showed how this level of assurance may determine the operations that people can perform. It demonstrated the control citizens have in practice over their virtual identities, allowing them to enrol with the ARIES ecosystem and build separate identities, for different purposes, effectively minimizing the disclosure of data and maximizing their privacy. This was informed by and designed to ensure implementation of ethical principles to help build trust.

The e-Commerce demonstrator scenario overview and main processes identified are shown in Figure 11.2. The demonstrator allowed users to use their own eID's present in the ARIES vault to register and to login using their biometrics (face authentication) on the Chef Continente website. This new authentication method was done in the mentioned website using the ARIES system and app to read real documents (e.g. passport) and biometrics (user's face) to validate identities and connect to the third-party e-Commerce service from Continente, the Portuguese leading retailer.

The demonstration stage aimed to validate ARIES' results in terms of applicability of the resulting ecosystem and enabling tools and technologies and effectively demonstrate the progress beyond the state of the art of ARIES achievements in a realistic scenario having potentially high impact on society

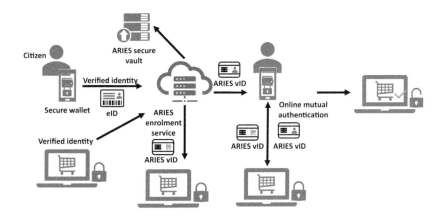

Figure 11.2 e-Commerce demonstrator scenario overview.

and the economy (this stage was done under strict ethical and legal requirements to protect participants). Several tools were used and are presented next in chronological order:

1. The ARIES Online Survey was designed to ascertain citizens' expectations of the ARIES eID and inform the project about which issues needed to be addressed (222 respondents);
2. Proof-of-Concept Design Thinking Workshop had the objective of to engage potential users with the ARIES system and get their feedback about relevancy, usability and functionalities (16 participants);
3. The Demonstrator Focus Groups stage had the main goal of testing the ARIES app in the real context with a group of users that had different backgrounds, ages and experiences (29 users).

The demonstrations were done amidst a time were the concern about privacy and security was at its highest, as this testing was coincidently done at the peak of a few international data breach scandals. Users stated that the focus on privacy, control of data and security was excellent and that they were exited that such a tool, might be available in the future. Also, the linkage of the project to EU funding generated an additional goodwill for it.

With regards to usability, users recognized that for a prototype the ARIES app had a very good look & feel, and the design was good, despite some minor usability issues, that were in the meantime solved. It's important to state that later versions, including the current one, included a more intuitive and visual explanation of the different steps and was recognized to be good.

Finally, users mentioned that the need for a passport as a baseline for the user data was an interesting approach as this could generate additional trust in the system. It was often mentioned that the control of own's data in the ARIES vault was very important and that the cloud updates on the data being reflected in the third-party services was a huge plus. All in all, it was with no surprise that most users stated that they would likely use the service if it was available in the market.

11.5.2 Airport Scenario Pilot and Validation

The planning of the airport demonstrator began in January 2018 the work focused on refining the platform functionality and defining "use cases", with the planning of logistics and final storyboards following as time progressed over the summer and into autumn. Over a period of months this helped establish the approximate date where the development of the ARIES technology and the availability of the venue would be at its most optimum.

The venue of the demonstrator was a crucial and unique location. It held challenges in bringing an innovative concept prototype into a realistic operational environment. The location chosen was the Leeds/Bradford International Airport at Yeadon, in the City of Leeds, West Yorkshire, England. This location chosen for the demonstrator pilot, had far greater challenges to overcome than normal locations this was due to the secure nature of the site and the need to occupy the airport on airside, to perform an operational test on the ARIES prototype in a controlled access environment.

During the consultation stage with the airport, Jet2.com, Border force and a retail outlet, identified that the optimum time during 2018 for holding the pilot was the month of November. This was when the airport had the least flights and passengers in the terminal building, staff availability was at its most convenient and it would minimise the disruption to normal business in the airport.

The end users and stakeholders who took part in the pilot were identified during the consultation period. They came from the airport, airline, and retail sections of the airline experience. In addition to take into account law enforcement participants came from counter terrorism, cyber and serious organised crime officers with also a focus on crime prevention and community cohesion.

All participants were first asked to perform a timed exercise where they performed the ARIES enrolment process to establish an eID using their own genuine passport. During this process a live capture image of their face is part of the enrolment process. This is compared to the biometric information held in the passport and is a verification of identity. Multiple identities were enrolled on the devices and by way of a password, each person could secure their personal data on the device.

Once this had been completed each participant was given an additional enrolment exercise to complete to demonstrate some of the functionality of the ARIES app. These included a genuine expired passport. The date on the passport had expired, so no enrolment could be completed. With a forged passport which was very noticeable, the participants were unable to complete enrolment. Using a stolen passport which was genuine but where none of the biometric information matched the participant was also tried, so no enrolment could take place and no creation of an eID in ARIES.

The pilot was completed in one day and covered two main themed functions in the airport namely the passenger gate boarding process and a retail shopping experience. To help demonstrate and test the security of the facial recognition technology and to maintain participant's engagement, they were asked to try to pass the boarding process wearing various head garments;

this served to obstruct the live capture, since a clear image of the face is required for comparison. The final section in the boarding scenario was a timed exercise. Using the four devices all the passengers were asked to line up in a queue and were then timed on four separate occasions going through the boarding process. To best simulate a queue of passengers who all have their own mobile devices, the participants were organized in a pre-defined group and asked to queue, in order that they could be rotated four times using the four mobile devices in rotation. It is possible in ARIES to simulate each passenger having their own device, by creating multiple different user profiles on one device. Once all the participants had completed the exercise they were asked to complete a feedback questionnaire.

The second phase of the pilot took place in the retail store. A laptop was set up on the cashier's desk to simulate both the register's screen and the customer's screen. A walk though demonstration was first performed for the retail manager and staff. During the demonstration, the staff were shown how ARIES could be used to present the information they required to approve a sale of a restricted item, such as alcohol or cigarettes. They were asked to consider that using a recognized vID provider could be an approved method of proof of identity. It was also explained that one of the objectives of the ARIES project is to protect customers' personal information and not to disclose unnecessary information about the customer that could be stolen and used in a fraudulent act. From their own mobile device, the customer has to consent to releasing the above personal data for the Cashier to view. They were then asked to complete a questionnaire.

The feedback questionnaires contained a set of generic qualitative based questions about the user experience during enrolment. Further sub-sections in the questionnaire focused on questions bespoke to the stakeholders involved, i.e. the passengers and airline's boarding experience and retailer's and customer's purchase experience. The questions also aimed to explore exploitation and marketability, in terms of how likely passengers/customers would be to use the app if it were available and how likely a business would be to exploit a product like ARIES. Once all the storyboards and questionnaires were completed, a final opportunity was given to all participants to ask questions and give any feedback not already covered in the questionnaire.

The airport demonstrator was designed to explore the effectiveness of ARIES in issuing virtual credentials in an operational environment which requires the highest level of assurance and eligibility. The pilot demonstrated that a virtual identity combined with a live capture would greatly increase the security that protects citizens' and their credentials with assurances to service providers and border officials, that the person is eligible to travel, or purchase items that are restricted by legislation.

The general performance of the prototype ARIES app and the verbal feedback given by participants, highlighted that the participants found the app easy to use and liked the concept of holding a duplicate electronic means of proving their identity; they also felt they were in control of that data.

Where ARIES failed to meet the KPI, was with the enrolment on the app; most participants commented that if they were enrolling at home rather than within a test environment, they would not have felt the pressure that came with performing the task in a timed session.

Comments from commercial end users focused on added security, reduced waiting times and efficiency savings in personnel. Comments also included the prospect of participants being able to spend more time in the commercial area of the airport, if the boarding process freed up waiting time. All participants saw clear benefits of the speedy boarding process; their customer experience in this stage of testing was very positive. While most users noted "some concern" in relation to concerns over privacy, no participants said they would not use the app. In fact, all users said if the app became a viable product, they would use it. Overall the pilot testing of the ARIES app was successful and the feedback useful and mostly positive.

11.6 Cyber-security and Privacy Research Challenges

The landscape of Identity Management (IDM) has been rapidly evolving since the effective launch of ARIES project in 2017. New identity management models have emerged, transcending the third party-based (identity provider) federated identity approach which has been dominating the identity and access management control landscape. In particular, multiple initiatives on Self-Sovereign Identity (SSI) are maturing and attracting attention from industry and governments [10].

These approaches are being supported on decentralised architectures enabled by Distributed Ledger Technologies (DLT) and more specifically Blockchain (noteworthy examples are Sovrin [11], Hyperledger Indy [14], uPort [12] or Blockstack [13]). Maturity of SSI solutions and widespread adoption has now a good basis on emerging international standards:

- W3C Credentials Community Group such as Decentralized Identifiers or DID [15]
- Decentralized Key Management System (DKMS) [16]

- DID Auth [17]
- Verifiable Credentials [18]

Furthermore, the European Commission contacted CEN/CLC, which has established a Focus Group on Blockchain and Distributed Ledger Technologies to collect identified European needs on these technologies, contextualised to Europe's specific normative and technological environment, monitoring relevant activities of the Joint Multi-Stakeholder Platform on ICT standardization and the Digitising European Industry initiative, while also supporting ISO/TC 307 with a possible future European Technical Committee on Blockchain and DLT, see [19].

The relevance of Law Enforcement Authorities for ARIES as key adopters engaged in the prevention and reduction of identity-related crimes, links well with analysed areas for use of Blockchain for Government and Public Services [20] and this points in the direction of more continuous development of ARIES sustainability when engaging with blockchain initiatives in the Public Sector, in particular, around the European Blockchain Forum/Observatory/Partnership [21] and future opportunities enabled by the development of a European Blockchain Services Infrastructure.

In this respect we can consider key technological breakthroughs achieved by ARIES and which define core features of its identity ecosystem which relate perfectly to major aspects required to materialise and achieve widespread adoption of the Self-Sovereign Identity paradigm:

1. ARIES implements a mobile wallet to manage derived identities, which is an essential client component in SSI approaches. This fully aligns with mobile identity orientation of ARIES, allowing full user-centric control of (user-owned) credentials and brings convenience and security to enrolment and authentication phases. In future phases, the ARIES app could also include an SSI Agent, acting as a trust anchor for establishing, by means of Agent-to-Agent Protocol, secure, authenticated connections to other agents (e.g. at relying parties). Coupled with DKMS protocol and Secure Element and Trusted Execution Element in mobile device, the wallet can maintain SSI private keys, extending use of these secure solutions already used for security of biometric material.

2. DID approach relates perfectly to ARIES approach of letting users manage multiple identities, bringing the additional advantage of allowing users to separate interactions and establish through DIDs encrypted channels with other entities (persons, organizations or things) to securely

exchange verifiable claims/credential data. This will allow to transition from an 'account-based' concept of IDM to based on user-managed connections over distributed blockchain solutions, with no central authorities that can be the target of attacks, thus achieving a more robust identity ecosystem.

3. ARIES derivation of reliable electronic identities from official or qualified credentials backed by the Member States (eIDASeID, ePassport) can be further explored, linking eIDAS network to ARIES provider for importing official identities into SSI infrastructure, see [22].

4. ARIES approach to data minimisation through Attribute Based Credentials, based on Zero Knowledge Proofs, aligns perfectly with the notion of SSI Verifiable Credentials, and allows once more, strict control by users of personal data disclosure with ease of use as cryptographic mechanisms ensure to relying parties that the user is in possession of certain attributes without revealing any additional unnecessary details, thus fulfilling GDPR data minimisation principle.

5. ARIES Secure Vault approach, with strict authorisation of access by users to competent identity crime investigation authorities, allows to explore in the future research possibilities to support this in private permissioned ledger technology, facilitating the cross-border investigation of identity related incidents and cooperation between Law Enforcerment Authorities.

All these aspects underline the readiness of ARIES results to reap opportunities, together with vibrant community of identity management experts, which are taking forward identity management ecosystems to a new paradigm of disintermediated and user-centric, privacy-respecting identity and access control. This will create clear benefits for the security of European citizens and organizations, helping authorities to collaboratively achieve EU strategic goals for identity fraud reduction.

11.7 Conclusion

As cyber-criminals evolve their cyber-attacks, the European Commission is determined to meet the challenge promoting research in different cyber-security and privacy areas to mitigate upcoming identity-related crimes in both virtual (i.e. misuse of information, cyber-mobbing) and physical places (i.e. people trafficking, organized crime). In this sense, during the last years, research efforts in different projects has been made to devise novel solutions aimed to increase user's privacy and protect them against evolving kind of cyber-crimes, the challenge is still ongoing.

To this aim, this chapter has summarized the main goals, challenges and the approach followed in European project H2020 ARIES project, whose ultimate goal is to provide security features that ensure highest levels of quality in secure credentials for highly secure and privacy-respecting physical and virtual identity management processes. In addition, the project has addressed key legal, ethical, socio-economic, technological and organisational aspects of identity-related crimes.

Novel processes such as virtual identity derivation, ACS, Id-proofings based on breeder documents, biometric process, along with security features (secure wallet, secure vaults) has been devised, implemented and validated for physical and virtual identity management, strengthening the link between physical-virtual identities to reduce identity fraud.

Acknowledgements

This book chapter received funding from the European Union's Horizon 2020 research and innovation programme under grant agreement No 700085 (ARIES project).

References

[1] The European Parliament, the Council of the European Union: Regulation (EU) no 910/2014 of the European parliament and of the council (2014).

[2] Hughes, J., Maler, E.: Security assertion markup language (saml) v2.0. Technical report, Organization for the Advancement of Structured Information Standards (2005).

[3] Hardt, D. (ed.): The oauth 2.0 authorization framework (2012).

[4] Camenisch, J., Van Herreweghen, E.: Design and implementation of the idemixanonymous credential system. In: Proceedings of the 9th ACM Conference on Computer and Communications Security, CCS 2002, pp. 21–30. ACM, New York (2002).

[5] Sabouri, A., Krontiris, I., Rannenberg, K.: Attribute-based credentials for trust (ABC4Trust). In: Fischer-Hübner, S., Katsikas, S., Quirchmayr, G. (eds.) TrustBus2012. LNCS, vol. 7449, pp. 218–219. Springer, Heidelberg (2012). doi:10.1007/978-3-642-32287-721

[6] Kortuem, G., Kawsar, F., Fitton, D., Sundramoorthy, V.: Smart objects as buildingblocks for the internet of things. IEEE Internet Comput. 14(1), 44–51 (2010).

[7] J. Camenisch and A. Lysyanskaya, "A signature scheme with efficient protocols," in Proc. Int. Conf. Secur. Commun. Netw., 2002, pp. 268–289.

[8] Jorge Bernal Bernabe, Jose L. Hernandez-Ramos, and Antonio F. Skarmeta Gomez, "Holistic Privacy-Preserving Identity Management System for the Internet of Things," Mobile Information Systems, vol. 2017, Article ID 6384186, 20 pages, 2017.

[9] J. L. C. Sanchez, J. Bernal Bernabe and A. F. Skarmeta, "Integration of Anonymous Credential Systems in IoT Constrained Environments," in *IEEE Access*, vol. 6, pp. 4767–4778, 2018. doi: 10.1109/ACCESS.2017.2788464

[10] European Blockchain Observatory, 'Blockchain innovation in Europe', https://www.eublockchainforum.eu/sites/default/files/reports/20180727_report_innovation_in_europe_light.pdf?width=1024&height=800&iframe=true

[11] Sovrin Foundation, https://sovrin.org/

[12] uPort, https://www.uport.me/

[13] Blockstack, https://blockstack.org/what-is-blockstack/

[14] Hyperledger Indy, https://www.hyperledger.org/projects/hyperledger-indy

[15] 'Decentralized Identifiers (DIDs) v0.11, Data Model and Syntaxes for Decentralized Identifiers (DIDs)', W3C Credentials Community Group Site: https://w3c-ccg.github.io/did-spec/

[16] DKMS, https://github.com/WebOfTrustInfo/rwot4-paris/blob/master/topics-and-advance-readings/ dkms-decentralized-key-mgmt-system.md

[17] 'Link to DID Auth final version', https://github.com/WebOfTrustInfo/rwot6-santabarbara/commit/c1c44d6d2ead845 db75f9a52b53c0fb4cd98 db2d

[18] W3C. Verifiable Credentials Working Group, https://www.w3.org/2017/vc/WG/

[19] 'Recommendations for Successful Adoption in Europe of Emerging Technical Standards on Distributed Ledger/Blockchain Technologies", ftp://ftp.cencenelec.eu/EN/EuropeanStandardization/Sectors/ICT/ Blockchain%20+%20DLT/FG-BDLT-White%20paper-Version1.2.pdf

[20] European Blockchain Observatory 'Blockchain for Government and Public Services', https://www.eublockchainforum.eu/sites/default/files/

reports/eu_observatory _blockchain_in_government_services_v1_2018-12-07.pdf?width=1024&height= 800&iframe=true

[21] European Blockchain Forum and Observatory, https://www.eublockchai nforum.eu/

[22] 'Importing National eID Attributes into a Decentralized IdM System', Abraham, A., June 2018,https://www.egiz.gv.at/files/projekte/2018/eId AttributeImport/ ImportNationaleEIdAttribute.pdf

12

The LIGHTest Project: Overview, Reference Architecture and Trust Scheme Publication Authority

Heiko Roßnagel[1] and Sven Wagner[2]

[1]Fraunhofer IAO, Fraunhofer Institute of Industrial Engineering IAO,
Nobelstr. 12, 70569 Stuttgart, Germany
[2]University Stuttgart, Institute of Human Factors and Technology
Management, Allmandring 35, 70569 Stuttgart, Germany
E-mail: heiko.roßnagel@iao.fraunhofer.de;
sven.wagner@iat.uni-stuttgart.de

There is an increasing amount of electronic transactions in business and peoples everyday lives. To know who is on the other end of the transaction, it is often necessary to have assistance from authorities to certify trustworthy electronic identities. The EU-funded LIGHTest project assists here, by building a global trust infrastructure using DNS, where arbitrary authorities can publish their trust information. This enables then an automatic verification process of electronic transactions. This paper gives an overview on the project, its reference architecture with its main components and its application fields.

12.1 Introduction

Traditionally, we often knew our business partners personally, which meant that impersonation and fraud were uncommon. Whether regarding the single European market place or on a Global scale, there is an increasing amount of electronic transactions that are becoming a part of peoples everyday lives, where decisions on establishing who is on the other end of the transaction is important. Clearly, it is necessary to have assistance from authorities to certify trustworthy electronic identities. This has already been done. For example, the EC and Member States have legally binding electronic signatures. But how can we query such authorities in a secure manner? With the current lack of a worldwide standard for publishing and querying trust information, this would be a prohibitively complex leading to verifiers having to deal with a high number of formats and protocols.

The EU-funded LIGHTest project attempts to solve this problem by building a global trust infrastructure where arbitrary authorities can publish their trust information. Setting up a global infrastructure is an ambitious objective; however, given the already existing infrastructure, organization, governance and security standards of the Internet Domain Name System, it is with confidence that this is possible. The EC and Member States can use this to publish lists of qualified trust services, as business registrars and authorities can in health, law enforcement and justice. In the private sector, this can be used to establish trust in inter-banking, international trade, shipping, business reputation and credit rating. Companies, administrations, and citizens can then use LIGHTest open source software to easily query this trust information to verify trust in simple signed documents or multi-faceted complex transactions.

The three-year LIGHTest project has started on September 1st, 2016 It is partially funded by the European Union's Horizon 2020 research and innovation programme under G.A. No. 700321. The LIGHTest consortium consists of 14 partners from 9 European countries and is coordinated by Fraunhofer-Gesellschaft. To reach out beyond Europe, LIGHTest attempts to build up a global community based on international standards and open source software.

The partners are ATOS (ES), Time Lex (BE), Technische Universität Graz (AU), EEMA (BE), Giesecke + Devrient (DE), Danmarks Tekniske Universitet (DK), TUBITAK (TR), Universität Stuttgart (DE), Open Identity Exchange (GB), NLnet Labs (NL), CORREOS (ES), University of Piraeus (GR), and Ubisecure (FI).

This paper provides on overview on the LIGHTest project, its reference architecture with its main components and its application fields. This overview is based on already published and accepted papers within this project. Due to the complexity and the wide-range of the project not all topics and work packages can be integrated in this paper. For more details, we refer to the LIGHTest project web site https://www.lightest.eu/.

This paper is structured as follows. Section 12.1 introduces related work. In Section 12.1 an overview of the LIGHTest reference architecture and usage scenarios examples are presented. The concept and role of the Trust Scheme Publication Authority (TSPA) is described in more detail in Section 12.1, by way of example for the components of the LIGHTest reference architecture. The TSPA is one of the key components of the LIGHTest reference architecture, which is used in every verification process. In Section 12.1, the Trust Police Language (TPL) and the Policy Authoring and Visualization Tools used in LIGHTest are introduced. A short discussion and outlook is given in Section 12.1 and a summary is provided in Section 12.1.

For further details, we refer to the following publications: [1] provided a first introduction into the LIGHTest project. In [2] the LIGHTest reference architecture and the Trust Scheme Publication Authority (TSPA) are presented. [3] proposes a delegation scheme that provides a general representation of delegations that can be extended to different domains. In [4] the external API of the involved components, and how they can be used to publish trust scheme information in the TSPA are described as well as how to use DNS to make trust scheme membership claims discoverable by a verifier in an automated way. If in addition to the Trust Scheme Membership, the requirements of the Trust Scheme are published, a Unified Data Model is required. In [5], the development and publication of such a Unified Data Model derived from existing trust schemes (e.g. eIDAS) is described. [6] present the Graphical Trust Policy Language (GTPL), as an easy-to-use interface for the trust policy language TPL proposed by LIGHTest. In [7], a low- and a high-fidelity prototype of the trust policy authoring tool were developed to evaluate the design, in particular considering novice users.

12.2 Related Work

Most of the existing trust infrastructures follow the subsidiarity principle. One prominent example is the eIDAS Regulation (EU) N° 910/2014 ([8]) on electronic identification and trust services for electronic transactions in the internal market. This includes that each Member State establishes and publishes national trusted lists of qualified trust service providers. For the access of these trusted lists, the EC publishes a central list ("List of Trusted Lists") which contains links to these lists. Due to the fact that for verifiers the direct use of trust lists can be very onerous, in particular for international electronic transactions, LIGHTest provides a framework that is conceptually comparable to OCSP for querying the status of individual certificates and which facilities the verification of trust.

DANE (DNS-based Authentication of Names Entities) is a standard using DNS and the DNS security extension DNSSEC to derive trust in TLS server certificates (RCF6698 [9] and RCF7218 [10]). For this purpose, the DNS resource record TLSA was introduced which associates a TLS server certificate (or public key) with the domain name where the record is found. Within LIGHTest, the DANE standard is used to secure network communication and where certificates are used for verifying data.

Much like TLSA, the SMIMEA mechanism [11] provides a number of ways to limit the certificates that are acceptable for a certain e-mail address. It associates an SMIME user's certificate with the intended domain name by certificate constraints. In LIGHTest, the SMIMEA resource record is used to verify if the certificate used for signing the trust list is valid.

For the publication that an entity operates under the trust scheme there is an existing and widely accepted standard for trust lists, which is ETSI TS 119 612 [12]. This standard provides "a format and mechanisms for establishing, locating, accessing and authenticating a trusted list which makes available trust service status information so that interested parties may determine the status of a listed trust service at a given time". Within LIGHTest, the ETSI TS 119 612 standard is used for the representation of Trust Lists.

12.3 Reference Architecture

This section gives an overview of the LIGHTest reference architecture. It defines the macroscopic design of the LIGHTest infrastructure as well as the overall system's components, their functionality and their interaction on a high-level view. Second, examples of usage scenarios are presented. For more details, we refer to [2].

12.3.1 Components of the Reference Architecture

Figure 12.1 shows the LIGHTest reference architecture with all the major software components and their interactions (see also [1] and [2]). It illustrates how a verifier can validate a received electronic transaction based on her individual trust policy and queries to the LIGHTest reference trust infrastructure.

The verifier interacts with the Policy Authoring and Visualization Tools (e.g. desktop or web applications). These tools also facilitate non-technical users the visualization and editing of trust policies, which can be individual

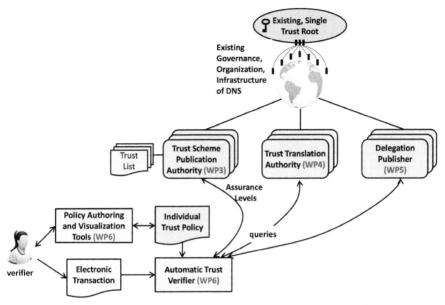

Figure 12.1 The LIGHTest reference architecture (see also [1, 2]).

and specific for each transaction. The role of the trust policy is the provision of formal instructions for the validation of trustworthiness for a given type of electronic transaction. For example, it states which trust lists from which authorities should be used. Further details are given in Section 12.1.

The Automatic Trust Verifier (ATV) takes the electronic transaction and trust policy as input and provides as output if the electronic transaction is trustworthy or not. In addition, the ATV may provide an explanation of its decision, in particular if the transaction was considered as not trustworthy.

The Trust Scheme Publication Authority (TSPA) uses a standard DNS Name Server with DNSSEC extension. A server publishes multiple trust lists under different sub-domains of the authority's domain name. The TSPA enables discovery and verification of trust scheme memberships. In Section 12.1, the TSPA is described in more detail.

The Trust Translation Authority also uses a standard DNS Name Server with DNSSEC extension. Here, a server publishes trust data under different sub-domains of the authority's domain name. In addition, trust translation lists express which authorities from other trust domains are trusted.

The Delegation Publisher uses a DNS Name Server with DNSSEC extension to discover the location (IP address) of the delegation provider, given that the user knows the correct domain name. The delegations themselves are not published in DNS mainly due to privacy reasons.

12.3.2 Usage Scenarios

In this section, examples of usage scenarios are presented. There are basic scenarios for trust publication, trust translation, and trust delegation, which can be used for qualified signatures, qualified seals, qualified identities, or qualified timestamps. The functionality (publish, translate, delegate) of the basic scenarios can be used to realise a wide range of more sophisticated scenarios. These scenarios can be either variants of the basic scenarios or a combination of different basic scenarios. A combination can be composing two trust services in a chaining process where the output level of the inner trust service becomes the input level of the outer trust service. For example, qualified delivery services, where E-registered delivery can be realised using a combination of the scenarios signature and timestamps. Another example is qualified website authentication, where trust publication with qualified identities is the basic scenario and additionally, trust translation could be used to e.g. authenticate third party users/things.

As an example for a basic scenario, a successful trust scheme membership verification for qualified signatures is presented. For this example, the following preconditions and assumptions for the electronic transaction and trust policy are made:

1. As preconditions, it is assumed that the verifier and signer are both located in the EC/eIDAS trust domain and that the eIDAS trust domain contains the actual eIDAS trust scheme. This means that trust translation is not required in this scenario. This could for example be managed in the following domain name structure: trust.ec.europa.eu - signature - TrustScheme - actual eIDAS trust scheme for qualified signature.

2. For the electronic transaction, it is assumed that the transaction is simply a signed document. Furthermore, the certificate used to sign the document contains a link to the trust list (Trust Membership Claim) for easier discovery such as "Issuer Alt Name: XYZ.qualified.trust.admin.ec" that points to the DNS resource records of the native trust scheme for qualified signatures. In addition, this trust scheme lists the certificate as qualified.

3. For the trust policy, it is assumed that trust policy simply states that the signature of the document is trusted if the issuer of the certificate is listed in TrustScheme.signature.trust.ec.europa.eu. Hence it is published as a

Boolean trust scheme publication (see Section 12.1 for the definition of Boolean trust scheme publication).

For the basic scenario of a successful trust scheme membership verification for qualified signatures with the preconditions and assumptions mentioned above, the corresponding information flow in the architecture is described in the following and depicted in Figure 12.2.

In step 1, the verifier feeds both, the Trust Policy and the Electronic Transaction into the ATV. The ATV parses the electronic transaction and yields the document, the signer certificate and the issuer certificate (step 2). In step 3, the ATV validates the signature on the document to make sure it is signed by the signer certificate. Next, the ATV validates that the signer certificate is signed by the issuer certificate (step 4). In step 5, the ATV searches the signer certificate and the issuer certificate for discovery information. The ATV finds a Trust Membership Claim in the signer certificate: "Issuer Alt Name: XYZ.qualified.trust.admin.ec". Hence, the issuer name is extracted from the certificate. In step 6, the ATV contacts the TSPA for retrieving the associated trust scheme. Therefore, the ATV issues a DNS query for all relevant resource records for boolean trust schemes for XYZ.qualified.trust.admin.ec. In step 7, the ATV verifies the chain of signatures from the DNS trust root of the

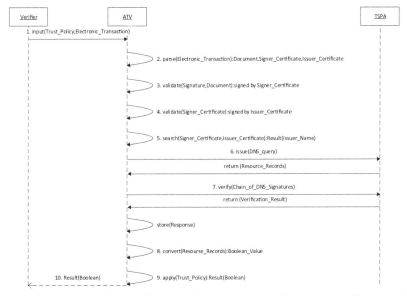

Figure 12.2 Sequence diagram for trust publication of a qualified signature (Boolean), [2].

DNS response using a validating resolver and stores the response as a "receipt" for future justification of its decision. Next, the ATV converts the resource records of the response into a boolean value (step 8). In the final step, the ATV looks at the trust policy and detects that the trust scheme, TrustScheme.signature.trust.ec.europa.eu is trusted (step 9). Hence, the overall result of applying the trust policy to the electronic transaction is trusted and sent back to the verifier (step 10).

The basic structure of the information flow for the other basic scenarios is similar. For qualified seals, qualified identities, or qualified timestamps it is mainly the domain name structure which differs. For trust translation, and trust delegation there are in addition some additional steps required using the Trust Translation Authority and the Delegation Publisher, respectively.

12.4 Trust Scheme Publication Authority

Knowing which trust scheme the issuer of the signers' certificate complies to is critical, in order to be able to verify whether an electronic transaction complies with the users' trust policy. It shows which security controls, and security requirements are fulfilled by the certificate issuer and thus indicate the security quality of the certificate that is used, e.g. for signing a document. The Trust Scheme Publication Authority (TSPA) is therefore an important component of the LIGHTest reference architecture. It enables discovery and verification of trust scheme memberships. Trust scheme publications are always associated with lists that indicate the membership of an entity with the referred to trust scheme. The described setup, which involve a trust list and a trust list provider aligns well with existing trust list standards (e.g. ETSI TS 119 612 [12]).

12.4.1 Trust Schemes and Trust Scheme Publications

A trust scheme itself can for example be constituted by requirements to information security processes, processes for issuance or revocation, requirements towards used technologies, or simply one single one-dimensional requirement, e.g. the geographical location of an entity. While some trust schemes, such as ETSI_EN_319_401 [13], just flatly lay out managerial requirements, trust schemes such as ISO/IEC 29115:2013 [14] further use different level of assurances to define which requirements must be met to comply with the trust scheme. In summary this all means, that a trust scheme can be published as a boolean trust scheme publication (e.g. [13]), and a

Table 12.1 Types of trust scheme publications in LIGHTest, [2]

Type of Trust Scheme Publication	Example	Verifiable Information
Boolean	ETSI_EN_319_401	Compliance of an entity to a trust scheme
Ordinal	LoA4.ISO29115	Compliance of an entity to an ordinal value of a trust scheme
Tuple-Based	{(authentication:2Factor), (identityProofing: inPerson)}	Requirements of a trust scheme

ordinal trust scheme publication (e.g. [14]) (see Table 12.1). Boolean trust scheme publications indicate the entities that comply with the requirements of the trust scheme, and thus are a member of the trust scheme. Ordinal trust scheme publications indicate the entities that comply with the requirements of an ordinal aspect (e.g. a level of assurance) of the trust scheme.

Both, Boolean and ordinal trust scheme publications do not provide any information on the requirements of the trust scheme, or the ordinal value (e.g. Level of Assurance) of the trust scheme that is represented by the trust scheme publication. In order to fill this gap, tuple-based trust scheme publications provide the requirements of a trust scheme in the form of attributes and values.

For this purpose, the development and publication of a unified Data Model derived from existing trust schemes (e.g. eIDAS) is needed, where each requirement is explicitly represented by one tuple. With this a unified view on the requirements of trust schemes is provided, which can be used within the TSPA. The consolidation and development of this Data Model, which is based on nine existing trust schemes, is presented along with possible applications in the field of trust verification in [5]. The unified Data Model includes the three abstract concepts Credential, Identity, and Attributes and in total 98 concepts, which can be added to standard Trust Lists using ETSI TS 119612.

12.4.2 Concept for Trust Scheme Publication Authority (TSPA)

The concept of the TSPA in LIGHTest consists of two components. It uses an off-the-shelf DNS Name Server with DNSSEC extension, in order to enable discovery of the Trust Scheme Provider that operates a Trust Scheme. The Trust Scheme Provider constitutes the second component of the TSPA. It provides a signed Trust List which indicates that a certificate Issuer is trusted under the scheme operated by the Trust Scheme Provider. It further provides the Tuple-Based representation of a Trust Scheme. As the DNS Name Server is only used to provide pointers to location of resources rather than storing the respective resources as DNS resource records directly, the TSPA is well-aligned with existing DNS practices. The use of pointers ensures the limited size of DNS messages, which is required for fast response times in the discovery process.

The use of the DNS Name Server system by LIGHTest enables easy and widespread adoption of the approach. We assume that the trust scheme of a certificate issuer is unknown, upon receiving an electronic transaction. The TSPA therefore provides the capability to discover a trust scheme member-ship claim for a certificate issuer, and verify this claim. The discovery of a trust scheme membership claim is done by using the domain name resolution capabilities of the DNS Name Server. Figure 12.3 provides an overview on the concept for trust scheme publishing in the TSPA. Since the TSPA is using the DNS Name Server mainly for pointing towards the Trust Scheme Provider and the tuple-based representation of a trust scheme, the concept is divided into the DNS records on the DNS Name Server (left side), and the data containers on the Trust Scheme Provider (right side).

The records on the DNS Name Server include a Data Container for the Issuer and for boolean and ordinal trust schemes. Data Containers for an Issuer are identified by an Issuer Name (indicated by *<IssuerName>*), and include the Name of the associated Trust Scheme. Data Containers for a Trust Scheme are identified by a SchemeName (indicated by *<SchemeName>*), in the boolean case, and an additional LevelName in the ordinal case (indicated by *<LevelName>.<SchemeName>*). A Trust Scheme data con-tainer includes the Trust Scheme Provider Domain Name (indicated by *<SchemeProviderName>*). The data containers for the Issuer, trust scheme name and ordinal level of a trust scheme include in addition certificate constraints, which enable to limit the certificates accepted for signing the trust list, using the SMIMEA DNS resource record. Hence, in the LIGHTest ecosystem, the SMIMEA resource record is used to verify if the certificate used for signing the trust list is valid. These records on the DNS Name Server have been developed in a consolidated approach to publishing trust-related information in general in the DNS within in LIGHTest project.

For the publication of tuple-based trust schemes, the tuples are published either in the signed trust list itself or listed in an extra document with a pointer from the signed trust list to this document. For both cases, there is

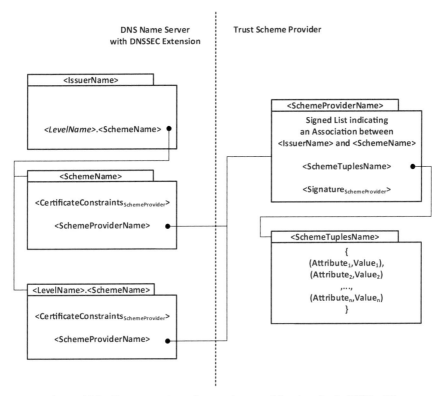

Figure 12.3 Representation of trust scheme publications in the TSPA, [5].

no additional DNS entry for the tuple-based trust schemes required. It uses the same as for the Trust Scheme Provider.

12.4.3 DNS-based Trust Scheme Publication and Discovery

The processes of Trust Scheme publication and discovery of trust lists using DNS is described in detail in [4]. To enable the automatic verification process of an electronic transaction using the ATV, it is required that the verifier knows where the trust scheme is saved at, and it would be more desirable if a CA can publish its membership claim. In order to be found in the DNS, each trust service and trust scheme taking part in LIGHTest picks a domain name as its identifier and announces this name in its associated certificates.

To update nameservers, the following two components were introduced: TSPA (concept of TSPA is introduced in Section 12.1) and ZoneManager.

The TSPA component itself acts as the endpoint for operators, which can be clients publishing trust schemes. It receives all relevant data via an HTTPS API to create the trust scheme. It can process links to existing trust schemes (e.g. eIDAS) as well as full trust scheme data. In the first case, the TSPA component creates the DNS entries together with the ZoneManager. In the second case, the TSPA component stores the trust scheme data locally and creates the DNS entries together with the ZoneManager. The second component, the ZoneManager, acts as the endpoint on the nameserver and modifies the zone data directly. It also ensures any zone data is properly signed using an existing DNSSEC setup. The ZoneManager's interface is only called from the TSPA component, and must never be called from the operator directly. Both components implement a RESTful API that is used by clients to publish the trust scheme information.

12.5 Trust Policy

As introduced in the Reference Architecture in Section 12.1, a verifier can validate a received electronic transaction based on her individual trust policy and queries to the LIGHTest reference trust infrastructure. To do so, the verifier has to provide the electronic transaction as well as an individual trust policy, which contains the formal instructions for the validation of trustworthiness for a given type of electronic transaction as input. The newly, in LIGHTest developed Policy Authoring and Visualization Tools facilitate and support also non-technical users to define their trust policies.

A Trust Policy is a recipe, expressed in a Trust Policy Language, that takes an Electronic Transaction and potentially multiple Trust Schemes, Trust Translation Schemes and Delegation Schemes as input and creates a single Boolean value (trusted [y/n]) and optionally an explanation (e.g., why not trusted) as output. For this purpose, a trust policy language is required, which is a formal language with well-defined semantics that is based on a mathematical formalism and is used to express the recipe of a trust policy. For the trust policy language in LIGHTest (LIGHTest TPL) the logic programming language Prolog that is based on Horn clauses only is used.

To facilitate the usage of LIGHTest TPL, [6] developed the Graphical Trust Policy Language (GTPL), which is an easy-to-use interface for the trust policy language TPL proposed by the LIGHTest project. GTPL uses a simple graphical representation where the central graphical metaphor is to consider the input like certificates or documents as forms and the policy author describes "what to look for" in these forms by putting constrains on the form's fields. GTPL closes the gap between languages on a logical-technical level such as TPL that require some expertise to use, and very basic interfaces like the LIGHTest Graphical-Layer that allow only a selection from a set of very basic patterns.

Furthermore, it is main goal of the project to develop and evaluate a trust policy authoring tool, considering especially novice users. As most contributions on usable policy authoring and IT-security only focus on the design phase of a tool and on stating guidelines how to make these tools and systems more user friendly. But there is a need for also evaluating tools, not only regarding usability but also user experience. For this purpose, a low- and a high-fidelity prototype were developed to evaluate the design (for further details see [7]). With the low-fidelity prototype a usability evaluation during the beginning of the design phase was conducted. After a design iteration a user experience evaluation with the high-fidelity prototype was conducted and the lessons learned derived from the results are considered.

12.6 Discussion and Outlook

The LIGHTest reference architecture and trust scheme publication authority (TSPA) support the implementation of the eIDAS Regulation ([8]). It enables the integration of existing trust lists using the global DNS infrastructure. Furthermore, it even expands eIDAS towards a global market and multi-users from the public and private sector. For the demonstration of the functionality of the LIGHTest infrastructure, two real world pilots are conducted within LIGHTest: In the first one, LIGHTest is integrated in the existing cloud based platform for trusted communication, the e-Correos platform. In the second one, LIGHTest is integrated in an existing e-Invoicing infrastructure and application scenario, OpenPePPOL.

Furthermore, key components of the LIGHTest infrastructure can be used for validation and authentication of data in sensor networks in IoT, e.g. for predictive maintenance use cases. This is demonstrated in a small sensor network of an organization using a Raspberry pi Cluster (see [15]).

LIGHTest supports UNHCR to explore ways to digitalize their documentation processes e.g. for the DAFI program. As the UNHCR deals with many sensitive documents and information, it is vital to be able to trust and verify the source of the documents after it is digitalized. This is especially important as it adds a higher level of security for such sensitive data and information. By digitalizing the documents using a Trust Scheme, it adds a level of security that not only optimizes the use of the digital documents, but also helps keep them secure. With that, after a trust scheme is made the digital documents created in the Trust Scheme can be verified and translated for both internal (with other UNHCR locations and Partners) or external (when the documents are being verified by other organizations that trust documents that are given to them by the UNHCR) purposes.

12.7 Summary

There is a high need for assistance from authorities to certify trustworthy electronic identities due to the worldwide increasing amount of electronic transactions. Within the EU-funded LIGHTest project, a global trust infrastructure based on DNS is built, where arbitrary authorities can publish their trust information. In this paper, a high level description of the LIGHTest reference architecture, its components and its application fields are presented. In addition, the Trust Scheme Publication Authority and the Trust Policy are described in more detail.

The reference architecture and the concept for Trust Scheme Publication Authority fulfil the main general principles and goals, which are required to develop a globally scalable trust infrastructure. Furthermore, it is well aligned with existing standards (e.g. ETSI TS 119 612) and fulfil the requirements using DNS name servers to build a global trust infrastructure.

In addition to the LIGHTest pilots for e-Correos and Open-PePPOL, there are a multitude of use cases, e.g. for sensor validation in the field of IoT or for international organizations (e.g. UNHCR).

Acknowledgements

This research is supported financially by the LIGHTest (Lightweight Infrastructure for Global Heterogeneous Trust Management in support of an open Ecosystem of Stakeholders and Trust schemes) project, which is partially funded by the European Union's Horizon 2020 research and innovation programme under G.A. No. 700321. We acknowledge the work and contributions of the LIGHTest project partners.

References

[1] Bruegger, B. P.; Lipp, P.: LIGHTest – A Lightweight Infrastructure for Global Heterogeneous Trust Management. In: Hühnlein D. et al.

(Hgeds.): Open Identity Summit 2016, Rome: GI-Edition, Lecture Notes in Informatics. S. 15—26.

[2] Wagner, S.; Kurowski, S.; Laufs, U.; Roßnagel, H.: A Mechanism for Discovery and Verification of Trust Scheme Memberships: The Lightest Reference Architecture. In (Fritsch, L.; Roßnagel, H.; Hühnlein, D., eds.): Open Identity Summit 2017. Gesellschaft für Informatik, Bonn, 2017.

[3] Wagner, G.; Omolola, O.; More, S.: Harmonizing Delegation Data Formats. In (Fritsch, L.; Roßnagel, H.; Hühnlein, D., eds.): Open Identity Summit 2017. Gesellschaft für Informatik, Bonn, 2017.

[4] Wagner, G.; Wagner, S.; More, S.; Hoffmann, H.: DNS-based Trust Scheme Publication and Discovery. In (Roßnagel, H.; Wagner, S.; Hühnlein, D., eds.): Accepted for Open Identity Summit 2019. Gesellschaft für Informatik, Bonn, 2019.

[5] Wagner, S.; Kurowski, S.; Roßnagel, H.: Unified Data Model for Tuple-Based Trust Scheme Publication. In (Roßnagel, H.; Wagner, S.; Hühnlein, D., eds.): Accepted for Open Identity Summit 2019. Gesellschaft für Informatik, Bonn, 2019.

[6] Mödersheim, S.; Ni, B.: GTPL: A Graphical Trust Policy Language. In (Roßnagel, H.; Wagner, S.; Hühnlein, D., eds.): Accepted for Open Identity Summit 2019. Gesellschaft für Informatik, Bonn, 2019.

[7] Weinhardt, S.; St. Pierre, D.: Lessons learned – Conducting a User Experience evaluation of a Trust Policy Authoring Tool. In (Roßnagel, H.; Wagner, S.; Hühnlein, D., eds.): Accepted for Open Identity Summit 2019. Gesellschaft für Informatik, Bonn, 2019.

[8] European Parliament, 'Regulation (EU) No 910/2014 of the European Parliament and of the Council of 23 July 2014 on electronic identification and trust services for electronic transactions in the internal market and repealing Directive 1999/93/EC', European Parliament, Brussels, Belgium, Regulation 910/2014, 2014.

[9] Hoffman, P.; Schlyter J.: The DNS-Based Authentication of Named Entities (DANE) Transport Layer Security (TLS) Protocol: TLSA, RFC 6698, DOI 0.17487/RFC6698, 2012, http://www.rfc-editor.org/info/rfc6698

[10] Gudmundsson, O.: Adding Acronyms to Simplify Conversations about DNS-Based Authentication of Named Entities (DANE), RFC 7218, DOI 10.17487/RFC7218, 2014, http://www.rfc-editor.org/info/rfc7218

[11] Hoffman, P.; Schlyter, J.: Using Secure DNS to Associate Certificates with Domain Names for S/MIME, RFC 8162, RFC Editor, May 2017.

[12] ETSI: Electronic Signatures and Infrastructures (ESI); Trusted Lists. Sophia Antipolis Cedex, France, Technical Specification ETSI TS 119 612 V1.1.1, 2013; http://www.etsi.org/deliver/etsi_ts/119600_119699/11 9612/01.01.01_60/ts_119612v010101p.pdf

[13] ETSI: Electronic Signatures and Infrastructures (ESI); General Policy Requirements for Trust Service Providers. ETSI, Sophia Antipois Cedex, France, European Standard ETSI EN 319 401, 2016; http://www.etsi.org/deliver/etsi_en/319400_319499/319401/02.01.01_60/en_31 9401v020101p.pdf

[14] ISO/IEC 29115: Information technology – Security techniques – Entity authentication assurance framework. ISO/IEC, Geneva, CH (2013).

[15] Johnson-Jeyakumar, I.-H.; Wagner, S.; Roßnagel, H.: Implementation of Distributed Light weight trust infrastructure for automatic validation of faults in an IOT sensor network. In (Rossnagel, H.; Wagner, S.; Hühnlein, D., eds.):Accepted for Open Identity Summit 2019. Gesellschaft für Informatik, Bonn, 2019.

13

Secure and Privacy-Preserving Identity and Access Management in CREDENTIAL

Peter Hamm[1], Stephan Krenn[2] and John Sören Pettersson[3]

[1]Goethe University Frankfurt, Germany
[2]AIT Austrian Institute of Technology GmbH, Austria
[3]Karlstad University, Sweden
E-mail: peter.hamm@m-chair.de; stephan.krenn@ait.ac.at;
john_soren.pettersson@kau.se

In an increasingly interconnected world, establishing trust between end users and service providers with regards to privacy and data protection is becoming increasingly important. Consequently, CREDENTIAL, funded under the European Union's H2020 framework programme, was dedicated to the development of a cloud-based service for identity provisioning and data sharing. The system aimed at offering both high confidentiality and privacy guarantees to the data owner, and high authenticity guarantees to the receiver. This was achieved by integrating advanced cryptographic mechanisms into standardized authentication protocols. The developed solutions were tested in pilots from three critical sectors, which proved that high user convenience, strong security, and practical efficiency can be achieved at the same time through a single system.

13.1 Introduction

Over the last decade, the availability and use of the Internet as well as the demand for digital services have massively increased. This demand has already reached critical and high assurance domains like governmental services, healthcare, or business correspondence. Those domains have particularly high requirements concerning privacy and security, as they are processing highly sensitive user data, and thus they need to be harnessed with various mechanisms for securing access.

Handling all the different authentication and authorization mechanisms requires user-friendly support provided by identity management (IdM) systems. However, such systems have recently experienced a paradigm shift themselves. While classical IdM systems used to be operated locally within organizations as custom-tailored solutions, nowadays identity and access management are often provided "as a service" by major cloud providers from different sectors such as search engines, social networks, or online retailers. Connected services can leverage the user identity base of such companies for authentication or identification of users.

In addition, many of these service providers do not only allow users to authenticate them towards a variety of cloud services, but also enable them to store arbitrary other, potentially sensitive, data on their premises, and share this data with other users in a flexible way, while giving the owner full control over who can access their data.

Unfortunately, virtually all existing solutions suffer from at least one of the following two drawbacks. Firstly, upon authentication a service provider (a.k.a. relying party) is only ensured by the IdP service that a user's attributes (e.g., name, birth data, etc.) are correct, but it does not receive any formal authenticity guarantees that these attributes were indeed extracted from, e.g., a governmentally-issued certificate. That is, the relying party needs to make assumptions about the trustworthiness of the IdP, which may not be desired in case of high-security domains. Secondly, users often do not get formal end-to-end confidentiality guarantees in the sense that the data storage and IdP do not have access to their data. In particular, for the IdP aspect this is technically necessary as otherwise the IdP could not vouch for the correctness of the claimed attributes. However, this introduces severe risks, e.g., in case of security incidents such as data leaks.

13.1.1 CREDENTIAL Ambition

The main ambition of the CREDENTIAL project was to overcome these limitations by designing and implementing a cloud-based identity and access management system which upholds privacy and data confidentiality at all times while simultaneously giving the relying party high and formal authenticity guarantees on the received data.

More precisely, the system aims to put users into full control over their data. They can share digitally signed data with relying parties in its entirety or in parts, thereby realizing the minimum disclosure principle.

Furthermore, all exchanged data is encrypted end-to-end, without the cloud-service provider being able to access the data. By being able to plausibly deny having access to the data, the service provider is able to build his business strategy around this advantageous security property. At the same time, the relying parties is guaranteed that the data they received from the identity provider is authentic and was indeed issues, e.g., by a public authority, thereby reducing the necessary amount of trust into the IdP with regards to the correctness of the provided data. This also holds true if only parts of a signed document are shared with the relying party.

13.2 Cryptographic Background

Before being able to describe how CREDENTIAL achieved its main ambition, we will briefly recap the necessary cryptographic primitives on a high level. For more detailed background information, we refer to the original literature.

13.2.1 Proxy Re-encryption

In conventional public key encryption schemes, a user Alice holds a public key pk_A and a corresponding secret key sk_A. Now, when another user Bob wants to send a message to Alice, he encrypts a message m under pk_A, and sends the resulting ciphertext c_A to Alice, who then can decrypt the ciphertext using her secret key. Unfortunately, this technique is not practical for data sharing applications: assume that Alice stores her confidential data in an encrypted form on a cloud platform. Now, in order to share the data with Bob and Charlie, she would need to download the ciphertexts, decrypt them locally, and encrypt them again under the right public keys, say pk_B and pk_C.

This challenge is overcome by proxy re-encryption, originally introduced by Blaze et al. [3], and later refined by Ateniese et al. [1] and Chow [6], among others. Using those schemes, Alice can use her secret key and a receiver's public key to compute a re-encryption key $rk_{A \rightarrow B}$. Using this key, a proxy can translate a ciphertext c_A encrypted for Alice into a ciphertext c_B for Bob, without learning any information about the message contained in the ciphertext beyond what is already revealed by the ciphertext itself (e.g., the size of the message).

Within CREDENTIAL, proxy re-encryption is used to enable end-to-end encrypted data sharing without negatively affecting usability or efficiency on the end-user side, as the computation is outsourced to the CREDENTIAL Wallet.

13.2.2 Redactable Signatures

Traditional digital signature schemes allow the receiver of a signed message to verify the authenticity of the document. That is, a signer first uses his secret signing key *sk* to sign a message *m*, obtaining a signature *sig*. Now, a receiver, having access to *m*, *sig*, and the signer's public verification key *vk* can verify that the message has not been altered in any way since the signature has been generated. In particular, any editing or deletion of message parts would be detected, as the verification process would fail.

While this is a very useful primitive in many applications, it is often too restrictive when developing privacy-preserving applications. For instance, when aiming for selective disclosure in authentication processes, the holder of a signed electronic identity document is not able to blank out the information he does not want to reveal to the receiver.

Redactable signatures [16] solve this problem. In such schemes, the signer can label blocks of a message *m* as admissible when creating a signature *sig*. Now, any party having access to *m* and *sig* can redact admissible message blocks and update the signature to a signature that will still verify for the altered message, without requiring any secret key material. However, no other modifications than redacting admissible blocks (such as deletion of other blocks or parts of blocks, or arbitrary updates to the messages) can be performed without breaking the validity of the signature. Thus, the receiving party can rest assured that the received data blocks are authentic and have been signed by the holder of the secret key.

13.3 Solution Overview

To realize the project's ambition, the project consortium developed a cloud-based platform called the CREDENTIAL Wallet. Users can access and manage their account using a mobile application, the CREDENTIAL App.

In the following, we describe the main steps performed by the actors involved in the CREDENTIAL authentication flow:

- A user obtains a digital certificate on his attributes from an issuer, which could be a public authority attesting the user's birth date or nationality, but also a service provider signing the expiration date of a subscription. This is done by letting the issuer sign the user's attributes using a redactable signature scheme.
- The user then encrypts the received certificate using his public encryption key and uploads this data to the CREDENTIAL Wallet.

- When a relying party – either another user or a service provider – requests access to the user's data for the first time, the user computes a re-encryption key from his public key to that of the relying party. To do so, the user employs the CREDENTIAL App, which fetches the receiver's public key, while the user's secret key is locally stored. The App then sends the re-encryption key to the CREDENTIAL Wallet, where it is stored in a dedicated key storage component. For subsequent access requests from the same relying party no fresh key material needs to be generated until a potential key update.
- Now, when the relying party accesses the data, the user receives a notification through the CREDENTIAL App. The user selects which attributes to reveal to the relying party and which ones to blank out. Having received the selection, the CREDENTIAL Wallet redacts the defined attributes and re-encrypts the resulting ciphertext for the receiver.
- Having received the re-encrypted and redacted data, the relying party decrypts the ciphertext using its own secret key and verifies the signature on the received attributes. If the verification succeeds, the receiver is ensured that the revealed information was indeed signed by the issuer, and continues, e.g., by granting the user access to the request resource. If the verification fails, authentication was unsuccessful and the relying party aborts.

An overview of the described data flow is given in Figure 13.1. In the case that a user wants to share non-authentic data with another user, the process is simplified, in the sense that all steps related to signature generation, redaction, and verification are omitted.

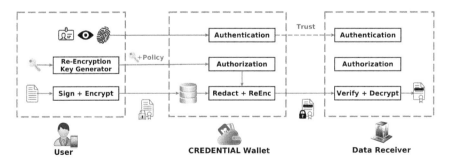

Figure 13.1   Abstract data flow in CREDENTIAL [10].

13.3.1 Added Value of the CREDENTIAL Wallet

The described data flow and implementation of the CREDENTIAL Wallet brings various benefits for all actors in the ecosystem of identity and access management [2, 11].

Benefits for end users. The end users of the CREDENTIAL Wallet benefit in various ways from the fact that the CREDENTIAL Wallet and all related components are under the privacy-by-design principle. For instance, the necessary trust into the IdP provider can be significantly reduced, as the provider does no longer have access to any of the user's data; besides protecting against internal threats such as malicious system administrators, this also shields the user against security incidents such as data leaks because of active attacks or during hardware decommissioning. Users are put back into full control over their data and can selectively disclose parts of their identity information to the service provider. This is enforced on a technical and not on a policy level. Furthermore, the user needs to access his or her secret key material when granting a relying party access to his attributes for the first time, but not for subsequent authentications. In particular, the user can store his or her secret key on a trusted mobile device, but does not need to carry it with them, e.g., when leaving for vacation. Finally, due to the implemented multi-factor authentication mechanisms, accessing a service from an insecure device (e.g., a shared PC) under the control of a potential adversary does not enable the adversary to impersonate the user for subsequent authentications to the same or other services.

Benefits for CREDENTIAL Wallet providers. Compared to traditional providers of identity and access management systems, providers of the CREDENTIAL Wallet benefit from the end-to-end encryption mechanisms used in our solution, and they can build their business models around our increased security features and guarantees. By not having access to sensitive user data, the liability risk is reduced significantly, and it becomes easier to comply with legal regulations such as the General Data Protection Regulations (GDPR).

Benefits for relying parties. The main benefit for relying parties is that they receive formal authenticity guarantees on the data they receive, by being able to verify that the data they receive was indeed cryptographically signed by a valid issuer. Consequently, they can significantly reduce the necessary trust into the identity provider. Furthermore, the CREDENTIAL Wallet was designed with maximum interoperability with existing industry standards for entity authentication (e.g., OAuth) in mind. This simplifies the integration into existing schemes substantially compared to other solutions following an ad-hoc design.

13.4 Showcasing CREDENTIAL in Real-World Pilots

A main objective of the CREDENTIAL project was not only to design the CREDENTIAL Wallet, improve and adapt the required technologies, and develop the necessary components, but also to evaluate the usability, stability, and efficiency of the applications in different real-world application domains from critical sectors.

In the following, we give a brief overview of the different pilot domains and our conclusions based on representative pilot users. Preliminary descriptions of the pilots can also be found in [8, 11].

13.4.1 Pilot Domain 1: eGovernment

CREDENTIAL's eGovernment pilot considered citizens and professionals who wish to authenticate themselves towards services offered by a public authority in a highly transparent way that gives them full control over which data goes where. More precisely, the project partners integrated the CREDENTIAL Wallet into SIAGE, a web portal hosted by our project partner Lombardia Informatica S.p.A. When visiting SIAGE's login page, users were offered to connect using their CREDENTIAL account. When selecting this option, they were redirected to an OpenAM component developed within the project, and an OAuth2 authentication flow was initiated. The users received a notification on their mobile phone and were asked to accept the information requested by the SIAGE system for authentication. Upon approval, the CREDENTIAL Wallet re-encrypted and redacted the appropriate user attributes before forwarding the resulting authentication token to SIAGE, which decrypted the data and verified its authenticity.

The pilot was executed using internal IT professionals for technical evaluations, and external focus groups to analyse the usability and perceived security aspects of the solution. The overall opinion of the users was very positive throughout all user groups. A detailed description of the pilot execution is also given in [17].

We want to stress that the analysed functionalities also demonstrate the technical feasibility and efficiency of the CREDENTIAL technologies in the context of many other eGovernment procedures beyond pure authentication, including aspects such as paper de-materialization. Imagine for example an employer who is willing to issue pay slips electronically. This employer, taking the role of the issuer in the authentication case, could sign the pay slip using a redactable signature scheme and label the different blocks of the pay slip as admissible. Now, when a user wants to request financial advantages from Lombardy region through the SIAGE system, he could log in as described above, and then decide to share those parts of the pay slip that are needed for receiving the requested support. For instance, if the support solely depends on gross income, the notification on the mobile phone would request obligatory access only to this data, and the user could decide to blank out information such as spent vacation days or reimbursements of actual travel costs. The data flows would be fully analogous to the authentication flow, and the service provider only needs to integrate the needed CREDENTIAL libraries.

13.4.2　Pilot Domain 2: eHealth

The eHealth pilot focused on secure remote data sharing between diabetes patients and their physicians [14]. To do so, two dedicated mobile applications for patients and doctors, respectively, have been developed.

The patient's app offers a convenient way for users to import medical data from devices such as glycosometers or scales. Like existing healthcare applications, users can browse through their history and get visual representations of their measurements. Whenever a user imports a new value and wishes to store it in its patient healthcare record (PHR), this access request is processed by a dedicated component developed within the project, cf. Figure 13.2. This so-called interceptor component redirects all requests through the CREDENTIAL Wallet. Technically, the patient's data is encrypted using a symmetric encryption scheme. The symmetric key is then encrypted employing the user's proxy re-encryption key and stored in their CREDENTIAL Wallet account. When selecting a treating doctor, the patient's application computes a re-encryption key from the patient to the doctor and deposits it in the CREDENTIAL Wallet's key store. Now, using the doctor's app, a diabetologist or general practitioner can access the encrypted key in the patient's account. The CREDENTIAL Wallet re-encrypts the ciphertext and the doctor receives the secret key that was used to encrypt the data in the PHR. After accessing the encrypted data in the PHR, the doctor can decrypt the data and analyse the patient's measurements.

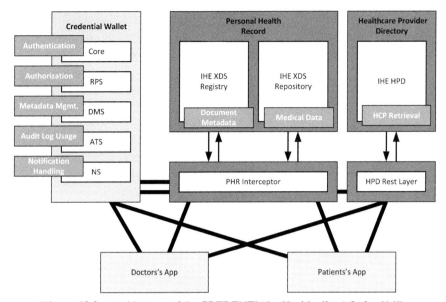

Figure 13.2 Architecture of the CREDENTIAL eHealth pilot (cf. also [14]).

Furthermore, the doctor can also provide feedback to the patient, e.g., by adding lab values such as HbA1c, or provide treatment recommendations.

After overcoming initial stability and efficiency problems, the feedback received from the external users and doctors was highly positive, in particular concerning the perceived security guarantees of the developed solution. One of our main conclusions is that it is possible to provide sophisticated end-to-end security solutions to the user in a way that is almost fully transparent and does not negatively affect usability.

13.4.3 Pilot Domain 3: eBusiness

The eBusiness pilot, documented in detail by Pallotti et al. [13], covered three use cases. The first use case allowed users to securely authenticate themselves towards an eCommerce platform, while the second use case enabled them to retrieve their data from the CREDENTIAL Wallet and share it with a service provider to subscribe to new services. From a technical point of view, these use cases are closely related to the eGovernment pilot described above, and we will focus on the third use case in the following.

In this use case, CREDENTIAL's proxy re-encryption libraries were integrated into InfoCert's Legalmail application, a certified mail service providing the same level of legal assurance as paper-based registered mail. The use case addressed the issue of forwarding encrypted emails to a deputy in case of absence: using classical email encryption technologies, the sender would need to be notified that the intended receiver is currently unavailable and would have to resend the mail to the defined deputy. The only way to avoid this additional interaction would then be to share the receiver's secret decryption key with the deputy, which however poses significant security risks and requires very high trust assumptions. Using proxy re-encryption, a Legalmail client can define a deputy, and deposit a re-encryption key at the mail server. Upon receiving an encrypted mail, the message is re-encrypted and forwarded to the deputy, who can decrypt the mail using his own secret key. While the sender does not need to be actively involved in this process, he still received a notification for transparency reasons.

The test users involved in the piloting phase showed genuine interest in the added security provided by CREDENTIAL. The possibility of exchanging confidential messages with a certified mail service has been highly appreciated. In addition, the pilot was able to show that the CREDENTIAL Wallet is not only able to provide meaningful features "as a whole", but also that single components of the system can successfully be integrated in other contexts.

13.5 Conclusion and Open Challenges

CREDENTIAL's main ambition was to develop a privacy-preserving and end-to-end secure and authentic data-sharing platform with integrated identity provisioning functionalities. To achieve this goal, the project consortium analysed, improved, and integrated security technologies from different domains including cryptography, multi-factor authentication, among others. Furthermore, the entire development process was accompanied by privacy experts to guarantee privacy-by-design and privacy-by-default, as well as by usability experts to ensure that end users are able to efficiently and conveniently interact with the system. Finally, the developed CREDENTIAL Wallet was tested through pilots within the highly sensitive domains of eGovernment, eHealth, and eBusiness, where the real-world usability and applicability of the developed solutions has been successfully proven.

13.5.1 Recommendations on Usability and Accessibility

Within the project, also ways to facilitate the adoption of privacy-friendly solutions for identity management and data sharing were studied. It turned out that users are often unaware of the privacy-issues with existing IdP solutions. Our analyses suggest that video tutorials can be an efficient way to inform users: statistical tests showed significant differences in the correctly identified advantages between participants who received a tutorial on single sign-on and those who did not, and also perceived usability increased of a more elaborate user interface which supported them in making more informed decisions [9].

Regarding accessibility, we believe that the European Directive 2016/2012 on the accessibility of websites and mobile applications of public sector bodies will inspire a development where assistive technology can seamlessly merge with IdP apps. A mobile application for the CREDENTIAL Wallet is an intermediary for the services benefitting from the Wallet's service. Thus, the public sector bodies – which all have to live up to the Directive – must rely on IdP services that also meet the requirements of the Directive. This will in its turn make it easy for service providers from the private sector to provide high levels of accessibility, as they benefit from users using these IdPs. Furthermore, the accessibility analysis provided within the CREDENTIAL project [7] can serve as an example for future developers of apps for services like the CREDENTIAL Wallet. One should also realise that further legal analysis might be needed from the public-sector side of its liabilities in accessible interactive communication.

13.5.2 Open Challenges

During the project duration, many challenges regarding design, efficiency, or understanding of user attitudes were successfully overcome. Nevertheless, we would like to briefly discuss two remaining challenges in the following.

Metadata privacy. From a technical point of view, metadata privacy is one of the main challenges that still needs to be addressed in cloud-based solutions such as the CREDENTIAL Wallet. While fundamental aspects such as linkability of authentication processes in cloud-based solutions were successfully addressed [12], the CREDENTIAL Wallet may still be able to infer sensitive information, e.g., who is sharing data with whom, or which data is accessed by whom and how often. The cryptographic literature contains several approaches to tackle these challenges, such as private information retrieval (cf. [4] and reference therein) or oblivious transfer (cf. [15] and the references given there). However, to the best of our knowledge, all existing solutions are currently too inefficient for large-scale deployment in real-world systems or would render the entire system too expensive.

Establishing business models for privacy. A major challenge we faced during the CREDENTIAL project relates to establishing sustainable business models for privacy-preserving solutions. At the current point in time, many major identity provider solutions – offered by, e.g., major search engine or social network providers – are free for the end user in the sense that no subscription fee needs to be paid, but the providers in turn gain substantial amounts of data about the user and build their business models around this. Furthermore, several studies have shown that while end users prefer privacy-preserving solutions in different scenarios, they are often not willing to pay for this feature. The successful commercialization of privacy-enhancing systems such as the CREDENTIAL Wallet would thus require a change of thinking on the cloud service provider and on the end user side, which could be triggered by legal regulations such as the General Data Protection Regulation (GDPR) or information campaigns to raise the users' awareness for privacy-related issues. Alternatively, especially for critical domains such as eHealth or eGovernment, we believe that also public authorities (e.g., ministry of health) could be potential providers of the CREDENTIAL Wallet, where the deployment and maintenance costs do not need to be paid directly by the end users.

Acknowledgements

The CREDENTIAL (Secure Cloud Identity Wallet) project leading to these results has received funding from the European Union's Horizon 2020 research and innovation programme under grant agreement No 653454.

The authors would like to thank all partners of the CREDENTIAL consortium for their efforts and work during the entire project duration and beyond. Finally, the authors are grateful to the editors for their efforts during the preparation of this book.

References

[1] Giuseppe Ateniese, Kevin Fu, Matthew Green, Susan Hohenberger. *Improved Proxy Re-Encryption Schemes with Applications to Secure Distributed Storage*. In NDSS 2005. The Internet Society. 2005.

[2] Charlotte Bäccman, Andreas Happe, Felix Hörandner, Simone Fischer-Hübner, Farzaneh Karegar, Alexandros Kostopoulos, Stephan Krenn, Daniel Lindegren, Silvana Mura, Andrea Migliavacca, Nicolas Notario McDonnell, Juan Carlos Pérez Baún, John Sören Pettersson, Anna E. Schmaus-Klughammer, Evangelos Sfakianakis, Welderufael Tesfay, Florian Thiemer, and Melanie Volkamer. *D3.1 – UI Prototypes v1*. CREDENTIAL Project Deliverable. 2017.

[3] Matt Blaze, Gerrit Bleumer, and Martin Strauss. *Divertible Protocols and Atomic Proxy Cryptography*. In EUROCRYPT 1998 (LNCS), Kaisa Nyberg (Ed.), Vol. 1403. Springer, 127–144. 1998.

[4] Ran Canetti, Justin Holmgren, Silas Richelson. *Towards Doubly Efficient Private Information Retrieval*. In TCC (2) 2017 (LNCS), Yael Kalai and Leonid Reyzin (Eds.), Vol. 10678. Springer, 694–726. 2017.

[5] Pasquale Chiaro, Simone Fischer-Hübner, Thomas Groß, Stephan Krenn, Thomas Lorünser, Ana Isabel Martínez Garcí, Andrea Migliavacca, Kai Rannenberg, Daniel Slamanig, Christoph Striecks, and Alberto Zanini. *Secure and Privacy-Friendly Storage and Data Processing in the Cloud*. In IFIP WG 9.2, 9.5, 9.6/11.7, 11.4, 11.6/SIG 9.2.2 International Summer School 2017 (IFIP AICT), Marit Hansen, Eleni Kosta, Igor Nai-Fovino, and Simone Fischer-Hübner (Eds.), Vol. 526. 153–170. 2018.

[6] Sherman S. M. Chow, Jian Weng, Yanjiang Yang, and Robert H. Deng. *Efficient Unidirectional Proxy Re-Encryption*. In AFRICACRYPT 2010 (LNCS), Daniel J. Bernstein and Tanja Lange (Eds.), Vol. 6055. Springer, 316–332. 2010.

[7] Felix Hörandner, Pritam Dash, Stefan Martisch, Farzaneh Karegar, John Sören Pettersson, Erik Framner, Charlotte Bäccman, Elin Nilsson, Markus Rajala, Olaf Rode, Florian Thiemer, Alberto Zanini, Alberto Miranda Garcia, Daria Tonetto, Anna Palotti, Evangelos Sfakianakis, and Anna Schmaus-Klughammer. *D3.2 – UI Prototypes v2 and HCI Patterns*. CREDENTIAL Project Deliverable. 2018.

[8] Felix Hörandner, Stephan Krenn, Andrea Migliavacca, Florian Thiemer, and Bernd Zwattendorfer. *CREDENTIAL: A Framework for Privacy-Preserving Cloud-Based Data Sharing*. In ARES. IEEE Computer Society, 742–749. 2016.

[9] Farzaneh Karegar, Nina Gerber, Melanie Volkamer, Simone Fischer-Hübner. *Helping John to make informed decisions on using social login*. In SAC 2018, Hisham M. Haddad, Roger L. Wainwright, and Richard Chbeir (Eds.). ACM, 1165–1174. 2018.

[10] Farzaneh Karegar, Christoph Striecks, Stephan Krenn, Felix Hörandner, Thomas Lorünser, and Simone Fischer-Hübner. *Opportunities and Challenges of CREDENTIAL - Towards a Metadata-Privacy Respecting Identity Provider*. In IFIP WG 9.2, 9.5, 9.6/11.7, 11.4, 11.6/SIG 9.2.2 International Summer School 2016 (IFIP AICT), Anja Lehmann, Diane Whitehouse, Simone Fischer-Hübner, Lothar Fritsch, and Charles D. Raab (Eds.), Vol. 498. 76–91. 2017.

[11] Alexandros Kostopoulos, Evangelos Sfakianakis, Ioannis Chochliouros, Jon Sören Pettersson, Stephan Krenn, Welderufael Tesfay, Andrea Migliavacca, and Felix Hörandner. *Towards the Adoption of Secure Cloud Identity Services*. In ARES. ACM, 90:1–90:7. 2017.

[12] Stephan Krenn, Thomas Lorünser, Anja Salzer, and Christoph Striecks. *Towards Attribute-Based Credentials in the Cloud*. In CANS 2017 (LNCS), Srdjan Capkun and Sherman S. M. Chow (Eds.), Vol. 11261. Springer, 179–202. 2018.

[13] Anna Pallotti, Luigi Rizzo, Romualdo Carbone, Pasquale Chiaro, and Daria Tonetto. *D6.6 – Test and Evaluation of Pilot Domain 3 (eBusiness)*. CREDENTIAL Project Deliverable. 2018.

[14] Anna Schmaus-Klughammer, Johannes Einhaus, and Olaf Rode. *D6.5 – Test and Evaluation of Pilot Domain 2 (eHealth)*. CREDENTIAL Project Deliverable. 2018.

[15] Peter Scholl. *Extending Oblivious Transfer with Low Communication via Key-Homomorphic PRFs*. In PKC (1) 2018 (LNCS), Michel Abdalla and Ricardo Dahab (Eds.), Vol. 10769. Springer, 554–583. 2018.

[16] Ron Steinfeld, Laurence Bull, and Yuliang Zheng. *Content Extraction Signatures*. In ICISC 2001 (LNCS), Kwangjo Kim (Ed.), Vol. 2288. Springer, 285–304. 2001.

[17] Alberto Zanini and Andrea Migliavacca. *D6.4 – Test and Evaluation of Pilot Domain 1 (eGovernment)*. CREDENTIAL Project Deliverable. 2018.

14

FutureTrust – Future Trust Services for Trustworthy Global Transactions

Detlef Hühnlein[1], Tilman Frosch[2], Jörg Schwenk[2],
Carl-Markus Piswanger[3], Marc Sel[4], Tina Hühnlein[1], Tobias Wich[1],
Daniel Nemmert[1], René Lottes[1], Stefan Baszanowski[1], Volker Zeuner[1],
Michael Rauh[1], Juraj Somorovsky[2], Vladislav Mladenov[2],
Cristina Condovici[2], Herbert Leitold[5], Sophie Stalla-Bourdillon[6],
Niko Tsakalakis[6], Jan Eichholz[7], Frank-Michael Kamm[7],
Jens Urmann[7], Andreas Kühne[8], Damian Wabisch[8], Roger Dean[9],
Jon Shamah[9], Mikheil Kapanadze[10], Nuno Ponte[11], Jose Martins[11],
Renato Portela[11], Çağatay Karabat[12], Snežana Stojičić[13],
Slobodan Nedeljkovic[13], Vincent Bouckaert[14], Alexandre Defays[14],
Bruce Anderson[15], Michael Jonas[16], Christina Hermanns[16],
Thomas Schubert[16], Dirk Wegener[17] and Alexander Sazonov[18]

[1]ecsec GmbH, Sudetenstraße 16, 96247 Michelau, Germany
[2]Ruhr Universität Bochum, Universitätsstraße 150, 44801 Bochum, Germany
[3]Bundesrechenzentrum GmbH, Hintere Zollamtsstraße 4, A-1030 Vienna, Austria
[4]PwC Enterprise Advisory, Woluwedal 18, Sint Stevens Woluwe 1932, Belgium
[5]A-SIT, Seidlgasse 22/9, A-1030 Vienna, Austria
[6]University of Southampton, Highfield, Southampton S017 1BJ, United Kingdom
[7]Giesecke & Devrient GmbH, Prinzregentestraße 159, 81677 Munich, Germany
[8]Trustable Limited, Great Hampton Street 69, Birmingham B18 6E, United Kingdom
[9]European Electronic Messaging Association AISBL, Rue Washington 40, Bruxelles 1050, Belgium
[10]Public Service Development Agency, Tsereteli Avenue 67A, Tbilisi 0154, Georgia

[11]Multicert – Servicos de Certificacao Electronica SA, Lagoas Parque Edificio 3 Piso 3, Porto Salvo 2740 266, Portugal

[12]Turkiye Bilimsel Ve Tknolojik Arastirma Kurumu, Ataturk Bulvari 221, Ankara 06100, Turkey

[13]Ministarstvo unutrašnjih poslova Republike Srbije, Kneza Miloša 103, Belgrade 11000, Serbia

[14]Arηs Spikeseed, Rue Nicolas Bové 2B, 1253 Luxembourg, Luxembourg

[15]Law Trusted Third Party Service (Pty) Ltd. (LAWTrust), 5 Bauhinia Street, Building C, Cambridge Office Park Veld Techno Park, Centurion 0157, South Africa

[16]Federal Office of Administration (Bundesverwaltungsamt), Barbarastr. 1, 50735 Cologne, Germany

[17]German Federal Information Technology Centre (Informationstechnikzentrum Bund, ITZBund), Waterloostr. 4, 30169 Hannover, Germany

[18]National Certification Authority Rus CJSC (NCA Rus), 8A building 5, Aviamotornaya st., Moscow 111024, Russia

E-mail: detlef.huhnlein@ecsec.de; tilman.frosch@rub.de; jorg.schwenk@rub.de; carl-markus.piswanger@brz.gv.at; marc.sel@be.pwc.com; tina.huhnlein@ecsec.de; tobias.wich@ecsec.de; daniel.nemmert@ecsec.de; rene.lottes@ecsec.de; stefan.baszanowski@ecsec.de; volker.zeuner@ecsec.de; michael.rauh@ecsec.de; juraj.somorovsky@rub.de; vladislav.mladenov@rub.de; cristina.condovici@rub.de; herbert.leitold@a-sit.at; sophie.stalla-bourdillon@soton.ac.uk; niko.tsakalakis@soton.ac.uk; jan.eichholz@gi-de.com; frank-michael.kamm@gi-de.com; jens.urmann@gi-de.com; kuehne@trustable.de; damian@trustable.de; r.dean@eema.org; jon.shamah@eema.org; mkapanadze@sda.gov.ge; nuno.pontemulticert.com; jose.martinsmulticert.com; renato.portelamulticert.com; cagatay.karabat@tubitak.gov.tr; snezana.stojicic@mup.gov.rs; slobodan.nedeljkovic@mup.gov.rs; vincent.bouckaert@arhs-developments.com; alexandre.defays@arhs-developments.com; bruce@LAWTrust.co.za; michael.jonas@bva.bund.de; christina.hermanns@bva.bund.de; thomas.schubert@bva.bund.de; dirk.dirkwegener@itzbund.de; sazonov@nucrf.ru;

Against the background of the regulation 2014/910/EU [1] on electronic identification (eID) and trusted services for electronic transactions in the internal market (eIDAS), the FutureTrust project[1], which is funded within the EU Framework Programme for Research and Innovation (Horizon 2020) under Grant Agreement No. 700542, aims at supporting the practical implementation of the regulation in Europe and beyond. For this purpose, the FutureTrust project will address the need for globally interoperable solutions through basic research with respect to the foundations of trust and trustworthiness, actively support the standardisation process in relevant areas, and provide Open Source software components and trustworthy services which will ease the use of eID and electronic signature technology in real world applications. The FutureTrust project will extend the existing European Trust Service Status List (TSL) infrastructure towards a "Global Trust List", develop a comprehensive Open Source Validation Service as well as a scalable Preservation Service for electronic signatures and seals. Furthermore, it will provide components for the eID-based application for qualified certificates across borders, and for the trustworthy creation of remote signatures and seals in a mobile environment. The present contribution provides an overview of the FutureTrust project and invites further stakeholders to actively participate in contributing to the development of future trust services for trustworthy global transactions.

14.1 Background and Motivation

There are currently over 160 trust service providers across Europe[2], which issue qualified certificates and/or qualified time stamps. Hence, the "eIDAS ecosystem"[3] with respect to these basic services is fairly well developed. On the other hand, the provision of qualified trust services for the validation and preservation of electronic signatures and seals as well as for registered delivery and the cross-border recognition of electronic identification schemes have been recently introduced with the eIDAS regulation [1]. However, these services are not yet broadly available in a mature, standardised, and interoperable manner within Europe.

In a similar manner, the practical adoption and especially the cross-border use of eID cards, which have been rolled out across Europe, is – despite

---

[1]See https://futuretrust.eu
[2]See [2, 3] and https://www.eid.as/tsp-map/ for example.
[3]See also https://blog.skidentity.de/en/eidas-ecosystem/.

previous and ongoing research and development efforts in pertinent projects, such as STORK, STORK 2.0, FutureID, e-SENS, SD-DSS, Open eCard, OpenPEPPOL and SkIDentity – still in its infancy. The opportunity afforded by the new eIDAS Trust Services regulation to use a national eID means outside of its home Member State, is still challenging and perceived to be complex. In particular, it is often not yet possible *in practice* to use eID cards from one EU Member State to enrol for a qualified certificate and qualified signature creation device (QSCD) in another Member State.[4]

In particular, the following problems seem to be not yet sufficiently solved and hence will be addressed in the FutureTrust project:

P1. No comprehensive Open Source Validation Service

Multiple validation services are available today. They range from offering revocation information to full validation against a formal validation policy. These services are operated by public and private sector actors, and allow relying parties the validation of signed or sealed artefacts. However, there is currently no freely available, standard conforming and comprehensive Validation Service, which would be able to verify arbitrary advanced and qualified electronic signatures in a trustworthy manner. To solve this problem, the FutureTrust project will contribute to the development of the missing standards and the development of such a comprehensive Validation Service.

P2. No scalable Open Source Preservation Service

The fact that signed objects lose their conclusiveness if cryptographic algorithms become weak induces severe challenges for applications, which require maintaining the integrity and authenticity of signed data for long periods of time. Research related to the strength of cryptographic algorithms is addressed in many places, including ECRYPT-NET[5], and does not fall within the scope of FutureTrust. Rather, the FutureTrust project will aim at solving this problem by contributing to the development of the missing standards for long-term preservation and the implementation of a scalable Open Source Preservation Service that makes use of processes and workflow to ensure preservation techniques embed the appropriate cryptographic solutions.

---

[4]Note, that such a cross-border enrolment for qualified certificates may become especially interesting in combination with remote and mobile signing services, in which no physical SSCD needs to be shipped to the user, because the SSCD is realized as central Hardware Security Module (HSM) hosted by a trusted service provider, which fulfils the requirements of [4], and against the background of the eIDAS-regulation (see e.g. Recital 51 of [1]) one may expect that such a scenario may soon become applicable across Europe and beyond.

[5]https://www.cosic.esat.kuleuven.be/ecrypt/net/

P3. Qualified electronic signatures are difficult to use in mobile environments

Today, applying for a qualified certificate involves various paper-based steps. Furthermore, to generate a qualified electronic signature, typically a smart card based signature creation device has to be used, which is complicated in mobile and cloud based environments due to the need for middleware and drivers that are often not supported on the mobile device. The FutureTrust project will aim at changing this by creating a Signature Service, which supports a variety of local and remote signature creation devices and eID-based enrolment for certificates and the remote creation of electronic signatures initiated by using mobile devices.

P4. Legal requirements of a pan-European eID metasystem

The first part of the eIDAS-regulation that deals with eID systems aims to create a standardized interoperability framework but does not intend to harmonize the respective national eID systems. Instead it employs a set of broad requirements, part of which is the mandatory compliance of all systems to the General Data Protection Regulation (GDPR) [5]. To facilitate compliance with the GDPR, the FutureTrust project will conduct desk research to analyse how privacy and data protection legislation impacts on existing laws and derive a list of necessary characteristics that an EU eID and eSignatures metasystem should incorporate to ensure compliance.

P5. Legally binding electronic transactions with non-European partners are hard to achieve

While the eIDAS-regulation [1] defines the legal effect of qualified electronic signatures, there is no comparable global legislation and hence electronic transactions with business partners outside the European Union are challenging with respect to legal significance and interoperability. To work on a viable solution for this problem the FutureTrust project will conduct basic research with respect to international legislation, contribute to the harmonization of the relevant policy documents and standards and build a "Global Trust List", which may form the basis for legally significant electronic transactions around the globe.

P6. Scope of eIDAS interoperability framework is limited to EU

In a similar manner, the scope of the interoperability framework for electronic identification according to Article 12 of [1] is limited to the EU. There are many aspects of an international interoperability framework that need to be assessed, especially in regard of to the privacy and data protection aspects highlighted above.[6] Against this background, the FutureTrust project will extend the work from pertinent research and large-scale pilot projects to integrate non-European eID-solutions in a seamless and trustworthy manner, after defining the requirements and assessing the impact of data transfers beyond the European Union.

P7. No formal foundation of trust and trustworthiness

To be able to compare eID solutions on an international scale, there is no international legislation which would allow to "define" trustworthiness. Instead, scientifically sound formal models must be developed which describe international trust models, and especially model to compare the trustworthiness of different eID services.

To demonstrate the viability and trustworthiness of these formal models, and show that the developed components can be used in productive environments, the FutureTrust project will implement real world pilot applications in the area of public administration, higher education, eCommerce, eBusiness and eBanking.

14.2 The FutureTrust Project

In order to solve the problems mentioned above, the FutureTrust partners (see Section 14.2.1) have sketched the FutureTrust System Architecture (see Section 14.2.2), which includes several innovative services, which are planned to be used in a variety of pilot projects (see Section 14.2.8).

This will in particular include the design and development of a Global Trust List (gTSL) (see Section 14.2.3), a Comprehensive Validation Service (ValS) (see Section 14.2.4), a scalable Preservation Service (PresS) (see Section 14.2.5), an Identity Management Service (IdMS) (see Section 14.2.6) and importantly a Signing and Sealing Service (SigS) (see Section 14.2.7).

14.2.1 FutureTrust Partners

The FutureTrust project is carried out by a number of core partners as depicted in Figure 14.1, which includes Ruhr-Universität Bochum (Germany), ecsec GmbH (Germany), Arhs Spikeseed (Luxembourg), EEMA

---

[6]For example, data transfers to the US are currently not clearly regulated after the invalidation of the 'Safe Harbor' agreement by the EUCJ (C-362/14). The EU officials were in negotiations on a new arrangement, named 'EU-US Privacy Shield' which was halted after a contradictory opinion from the WP29 (WP238).

Figure 14.1 FutureTrust partners.

(Belgium), Federal Computing Centre of Austria (Austria), Price Waterhouse
Coopers (PWC) (Belgium), University of Southampton (United Kingdom),
multicert (Portugal), Giesecke & Devrient GmbH (Germany), Trustable
Ltd. (United Kingdom), Secure Information Technology Center – Austria
(Austria), Public Service Development Agency (Georgia), Türkiye Bilimsel
veTeknolojik Araşrma Kurumu (Turkey), LAW Trusted Third Party Services
(Pty) Ltd. (South Africa), Ministry of Interior Republic of Serbia (Serbia),
DFN-CERT Services GmbH, the PRIMUSS cluster consisting of ten Univer-
sities of Applied Science and the Leipzig University (LU) Computing Centre
(Germany).

Furthermore the FutureTrust project is supported by selected subcon-
tractors and a number of associated partners, which currently includes the
SAFE Biopharma Association (USA), The Data Processing Center (DPC)
of the Ministry of Transport, Communications and High Technologies of
the Republic of Azerbaijan, Signicat, SK ID Solution AS, B.Est Solutions,
UITSEC Teknoloji A.Ş., and Comsign Israel.

14.2.2 FutureTrust System Architecture

As shown in Figure 14.2, the FutureTrust system integrates existing and
emerging eIDAS Trust Services, eIDAS Identity Services and similar Third

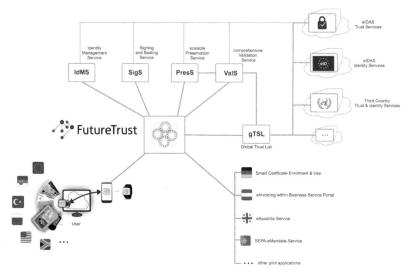

Figure 14.2 FutureTrust System Architecture.

Country Trust & Identity Services and provides a number of FutureTrust specific services, which aim at facilitating the use of eID and electronic signature technology in different application scenarios.

14.2.3 Global Trust List (gTSL)

The gTSL will become an Open Source component, which can be deployed with the other FutureTrust services or as standalone service and which allows to manage Trust Service Status Lists for Trust Services and Identity Providers. The gTSL will allow to import the European "List of the Lists" (LotL), which is a signed XML document according to [6] and all national Trust Service Status Lists (TSLs) referenced therein. This LotL is currently published by the European Commission. This import includes a secure verification of the digital signatures involved. The gTSL will also allow to import Trusted Lists from other geographic regions, such as the Trust List of the Russian Federation[7] for example, and it is envisioned that the gTSL will generate a "virtual US-American Trust List" from the current set of available cross-certificates. gTSL will provide support for the traceable assessment of trust related aspects for potential trust anchors both with and

---

[7]See http://e-trust.gosuslugi.ru/CA/DownloadTSL?schemaVersion=0.

without known trustworthiness and assurance levels[8] by providing claims or proofs of relevant information with respect to the trustworthiness of a trust service. This may give rise for a reputation based "web of trust" for trust services. It is expected that the corroboration of information from relatively independent sources[9] will help to establish trustworthiness. Furthermore, the gTSL provides a web interface as well as a REST interface allowing for a small set of predefined queries, to allow the other FutureTrust services or other gTSL deployments to access the validated data. For implementation of the underlying gTSL model various options have already been identified. These include traditional models such as a Trusted Third Party model and a Trust List, as well as innovative models such as a semantic web ontology and a blockchain ledger.

14.2.4 Comprehensive Validation Service (ValS)

The major use case of ValS is the validation of Advanced Electronic Signatures (AdES) in standardized formats, such as CAdES, XAdES and PAdES for example. In order to support the various small legal and regulatory differences with respect to electronic signatures coming from different EU Member States or other global regions, the ValS will support practice oriented XML-based validation policies for electronic signatures, which consider previous work in this area, such as [7] and [8] and current standards, such as [9] and [10] for example. The ValS issues a verification report to the requestor of the service, which is based on the recently published ETSI TS 119 102-2 signature validation report, which in particular considers the procedures defined in [9] and the XML-based validation policies mentioned above. Finally, it seems worth to be mentioned that the ValS is designed in a modular and extensible manner, such that modules for other not (yet) standardized signatures or validation policies can be plugged into the ValS in a well-defined manner.

14.2.5 Scalable Preservation Service (PresS)

The PresS is used to preserve the integrity and conclusiveness of a signed document over its whole lifetime. For this purpose the FutureTrust

---

[8][1] implicitly defines the levels "qualified" and "non-qualified" for trust service providers and explicitly introduces in Article 8 the assurance levels "low", "significant" and "high" for electronic identification schemes.

[9]See [11].

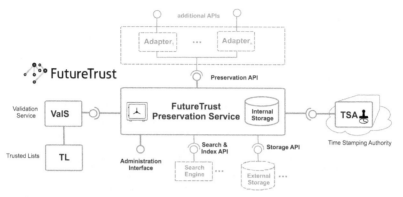

Figure 14.3   Outline of the Architecture of the Scalable Preservation Service.

Preservation Service as outlined in Figure 14.3 will use the ValS and existing external time stamping services to produce Evidence Records according to [12]. As depicted in Figure 14.3 the Preservation Service supports the input interface, which is currently standardised in ETSI TS 119 512 and smoothly integrates with various types of storage systems.

The FutureTrust Preservation Service will support a variety of Archive Information Packages including the zip-based container based on the Associated Signature Container (ASiC) specification according to [13]. An important goal of the envisioned Preservation Service is scalability, which may be realized by using efficient data structures, such as Merkle hash trees as standardized in [12] for example. Using hash tree based signatures[10] may also provide additional security in the case that quantum computers have been built, because any digital signature that is in use today (based on the RSA assumption or on the discrete log assumption) can be forged in this case. However, message authentication codes (MACs), block-chain constructions and signature algorithms based on hash-trees seem to remain secure. Thus it is an interesting research question, whether fully operational and sufficiently performant preservation services can be built on MACs, block-chains or hash-trees alone.

14.2.6 Identity Management Service (IdMS)

Many EU Member States and some non-European countries have established eID services, which produce slightly different authentication tokens. Within

---

[10]See [14].

the EU, most[11] of these services produce SAML tokens (see [15]) and the eIDAS interoperability framework [16] is also based on [17]. In addition, industrial standardization activities have produced specifications like FIDO[12] or GSMA's MobileConnect[13] which have gained a broad customer base. The IdMS is based on SkIDentity [18] and is able to consume a broad variety of such authentication tokens (SAML, OpenID Connect, OAuth), work with a broad variety of mobile identification services (FIDO, GSMA Mobile-Connect, European Citizen cards) and transform them into a standardized, interoperable[14] and secure[15] format. The choice of this standardized format will be based on industry best practices, and on the eIDAS interoperability framework [16]. Moreover, the IdMS supports a large variety of European and non-European eID cards, platforms and application services.

14.2.7 Signing and Sealing Service (SigS)

The SigS allows to create advanced and qualified electronic signatures and seals using local and remote signature generation devices. For this purpose, the SigS is operated in a secure environment and supports appropriate standard interfaces based on OASIS DSS-X Version 2.

As outlined in Figures 14.4 and 14.5, one may distinguish the enrolment phase and the usage phase. During enrolment, the Signatory uses his eID and the IdMS to perform an eID-based identification and registration at the SigS and the Certification Authority (CA), which involves the creation of signing credentials, which can later on be used for signature generation. Thanks to the

---

[11]The [19] system seems to be an exception to this rule, as it produces and accepts identity tokens according to the [20] specification.

[12]See [21].

[13]See [22] and [23].

[14]Due to the fact that SAML is a very complex and highly extensible standard, the integration of different eID services considering all extensions points is a rather challenging task. In order to enable the communication between all eID services, their interoperability has to be thoroughly analysed.

[15]Based on [16] it is clear that SAML 2.0 will form the basis for eIDAS Interoperability Framework according to Article 12 of [1] and [24], but it is currently likely that the Assertions will be simple "Bearer Tokens", which is not optimal from a security point of view. Furthermore, the different authentication flows and optional message encryptions result in complex standard and thus expose conforming implementations to new attacks. In the last years, several papers (see e.g. [25]) showed how to login as an arbitrary use in SAML Single Sign-On scenarios or decrypt confidential SAML messages (see e.g. [26]). Thus, existing eID services can be evaluated against known attacks.

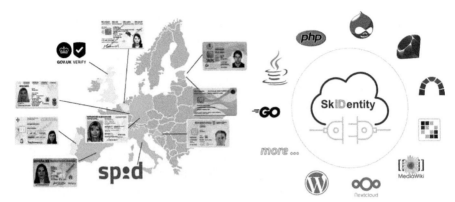

Figure 14.4   National eID cards, platforms and applications supported by IdMS.

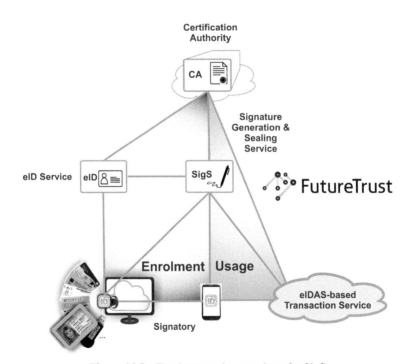

Figure 14.5   Enrolment and usage phase for SigS.

OASIS DSS Extension for Local and Remote Signature Computation [27] it is possible to use both smart card and cloud-based signature creation devices.

14.2.8 FutureTrust Pilot Applications

The FutureTrust consortium aims to demonstrate the project's contributions in a variety of demonstrators and pilot applications, which are planned to include University Smart Certificates Enrolment & Use, e-Invoicing with the Business Service Portal of the Austrian Government, an e-Apostille Validation System and a SEPA e-Mandate Service according to [28] for example. Furthermore, the FutureTrust project is open for supporting further pilot applications related to innovative use cases for eID and electronic signature technology.

14.2.9 The go.eIDAS Initiative

It is recognised that the FutureTrust Service components that will be made available exist in the eIDAS ecosystem and all exploitation efforts must reflect the early stages of Trust Services deployment and market maturity. In order to establish FutureTrust to be sustainable and to maintain its relevance, it is essential to obtain the best possible support for the exploitation efforts, especially from others than the FutureTrust Partners, Associate Partners and Advisory Board Members. To this end, the go.eIDAS[16] initiative will act as the exploitation vehicle for FutureTrust, but will also have sufficient branding to continue after the end of the Horizon 2020 funding.

Planning and initial contacts with Stakeholders commenced with the launch press release on 27/09/2018, in conjunction with the formal start of EU recognition of notified eIDs[17].

go.eIDAS reflects the private sector need to interoperate with eIDAS and also to interoperate with non-EU based Trust Schemes. go.eIDAS is an open initiative, which welcomes all interested organisations and individuals who are committed to the goals of eIDAS and FutureTrust. We recognise that a thriving community with a spectrum of needs must be created over and above the users of FutureTrust.

---

[16] See https://go.eid.as

[17] See https://www.eid.as/news/details/date/2018/09/27/goeidas-initiative-launched- across-europe-and-beyond-1/.

14.3 Summary and Invitation for Further Collaboration

This paper provides an overview of the FutureTrust project, which started on June 1st 2016 and is funded until August 2019 by the European Commission within the EU Framework Programme for Research and Innovation (Horizon 2020) under the Grant Agreement No. 700542 with up to 6,3 Mio. €.

As explained throughout the paper, the FutureTrust project has conducted basic research with respect to the foundations of trust and trustworthiness, actively support the standardisation process in relevant areas, and plans to provide innovative Open Source software components and trustworthy services which will enable ease the use of eID and electronic signature technology in real world applications by addressing the problems P1 to P7 introduced in Section 14.2.

As part of the continuation of this project, and its subsequent exploitation, the FutureTrust consortium invites interested parties, such as Trust Service Providers, vendors of eID and electronic signature technology, application providers and other research projects to benefit from this development and join the FutureTrust team in its new go.eIDAS initiative.

References

[1] 2014/910/EU. (2014). Regulation (EU) No 910/2014 of the European Parliament and of the council on electronic identification and trust services for electronic transactions in the internal market and repealing Directive 1999/93/EC. http://eur-lex.europa.eu/legal-content/EN/TXT/?uri=uriserv:OJ.L_.2014.257.01.0073.01.ENG.

[2] EU Trusted Lists of Certification Service Providers. (2016). *European Commision.* Retrieved from https://ec.europa.eu/digital-agenda/en/eu-trusted-lists-certification-service-providers

[3] 3×A Security AB. (2016). EU Trust Service status List (TSL) Analysis Tool. http://tlbrowser.tsl.website/tools/.

[4] CEN/TS 419 241. (2014). Security Requirements for Trustworthy Systems supporting Server Signing.

[5] 2016/679/EU. (2016). Regulation (EU) 2016/679 of the European Parliament and of the Council of 27 April 2016 on the protection of natural persons with regard to the processing of personal data and on the free movement of such data, *and repealing Directive 95/46/EC and repealing Directive 95/46/EC (General Data Protection*

Regulation) (Text with EEA relevance). http://eur-lex.europa.eu/legal-content/EN/TXT/?uri=uriserv:OJ.L_.2016.119.01.0001.01. ENG&toc= OJ:L:2016:119:TOC.

[6] ETSI TS 119 612. (2016, April). Electronic Signatures and Infrastructures (ESI); Trusted Lists. *Version 2.2.1*. http://www.etsi.org/deliver/etsi_ts/119600_119699/119612/02.02.01_60/ts_ 119612v020201p.pdf.

[7] ETSI TR 102 038. (2002, April). TC Security - Electronic Signatures and Infrastructures (ESI); XML format for signature policies.

[8] ETSI TS 102 853. (2012, July). Electronic Signatures and Infrastructures (ESI); Signaturue verification procedures and policies. *V1.1.1*. http://www.etsi.org/deliver/etsi_ts/102800_102899/102853/01.01.01_60/ts_ 102853v010101p.pdf.

[9] ETSI EN 319 102-1. (2016, May). Electronic Signatures and Infrastructures (ESI); Procedures for Creation and Validation of AdES Digital Signatures; Part 1: Creation and Validation, Version 1.1.1. http://www.etsi.org/deliver/etsi_en/319100_319199/31910201/01.01.01_60/en_ 31910201v010101p.pdf.

[10] ETSI TS 199 172-1. (2015, July). Electronic Signatures and Infrastructures (ESI); Signature Policies; Part 1: Building blocks and table of contents for human readable signature policy documents. http://www.etsi.org/deliver/etsi_ts/119100_119199/11917201/01.01.01_60/ts _11917201v010101p.pdf.

[11] Sel, M. (2016). Improving interpretations of trust claims, published in the proceedings of the. *Trust Management X: 10th IFIP WG 11.11 International Conference, IFIPTM 2016* (pp. 164–173). Darmstadt, Germany: Springer.

[12] RFC 4998. (2007, August). Gondrom, T.; Brandner, R.; Pordesch, U. *Evidence Record Syntax (ERS)*. https://tools.ietf.org/html/rfc4998.

[13] ETSI EN 319 162-1. (2015, August). Electronic Signatures and Infrastructures (ESI); Associated Signature Containers (ASiC); Part 1: Building blocks and ASiC baseline containers. http://www.etsi.org/deliver/etsi _en/319100_319199/31916201/01.00.00_20/en_ 31916201v010000a.pdf.

[14] Buchmann, J., Dahmen, E., and Szydlo, M. (2009). Hash-based digital signature schemes. In *Post-Quantum Cryptography* (pp. 35–93). Springer.

[15] Zwattendorfer, B., & Zefferer, T. T. (2012). The Prevalence of SAML within the European Union. *8th International Conference on Web Information Systems and Technologies (WEBIST)*, (pp. 571–576). http://www.webist.org/?y=2012.

[16] eIDAS Spec. (2015, November 26). eIDAS Technical Subgroup. *eIDAS Technical Specifications v1.0.* https://joinup.ec.europa.eu/software/cefe id/document/eidas-technical-specifications-v10.

[17] SAML 2.0. (2005, March 15). *OASIS Standard.* Retrieved from Metadata for the OASIS Security Assertion Markup Language (SAML) V2.0: http://docs.oasis-open.org/security/saml/v2.0/saml-metadata-2.0-os.pdf

[18] SkIDentity. (2018). Retrieved from https://www.skidentity.com/

[19] FranceConnect. (2016). https://doc.integ01.dev-franceconnect.fr/.

[20] OpenID Connect. (2015). OpenID Foundation. *Welcome to OpenID Connect.* http://openid.net/connect/.

[21] FIDO. (2015). *FIDO Alliance.* Retrieved from https://fidoalliance.org/

[22] GSMA. (2015). *Introducing Mobile Connect – the new standard in digital authentication.* Retrieved from http://www.gsma.com/personaldata/mobile-connect

[23] GSMA-CPAS5. (2015). CPAS 5 OpenID Connect - Mobile Connect Profile - Version 1.1. https://github.com/GSMA-OneAPI/Mobile-Conn ect/tree/master/specifications.

[24] 2015/1501/EU. (2015). Commission Implementing Regulation (EU) 2015/1501 of 8 September 2015 on the interoperability framework pursuant to Article 12(8) of Regulation (EU) No 910/2014. *of the European Parliament and of the Council on electronic identification and trust services for electronic transactions in the internal market (Text with EEA relevance).* http://eur-lex.europa.eu/legal-content/EN/TXT/?uri=OJ%3AJOL_2015_235_R_0001.

[25] Somorovsky, J., Mayer, A., Schwenk, J., Kampmann, M., & Jensen, M. (2012). On breaking saml: Be whoever you want to be. *Presented as part of the 21st USENIX Security Symposium (USENIX Security 12).*

[26] Jager, T., & Somorovsky, J. (2011). How to break xml encryption. *Proceedings of the 18th ACM conference on Computer and communications security.*

[27] OASIS DSS Local & Remote Signing (2018). *DSS Extension for Local and Remote Signature Computation Version 1.0.*

[28] EPC 208-08. (2013, April 9). European Payments Council. *EPC e-Mandates e-Operating Model - Detailed Specification.* Version 1.2: http://www.europeanpaymentscouncil.eu/index.cfm/knowledge-bank/ep c- documents/epc-e-mandates-e-operating-model-detailed-specification/ epc208-08-e-operating-model-detailed-specification-v12-approvedpdf/.

[29] ETSI SR 019 020. (2016, February). The framework for standardisation of signatures: Standards for AdES digital signatures in mobile and distributed environments. *V1.1.1*. http://www.etsi.org/deliver/etsi_sr/0190 00_019099/019020/01.01.01_60/sr_ 019020v010101p.pdf.

[30] ETSI TR 119 000. (2016, April). Electronic Signatures and Infrastructures (ESI); The framework for standardization of signatures: overview. Version 1.2.1: http://www.etsi.org/deliver/etsi_tr/1190 00_119099/119000/01.02.01_60/tr_ 119000v010201p.pdf.

[31] Kubach, M., Leitold, H., Roßnagel, H., Schunck, C. H., & Talamo, M. (2015). SSEDIC.2020 on Mobile eID. *to appear in proceedings of Open Identity Summit 2015.*

[32] Kutylowski, M., & Kubiak, P. (2013, May 06). Mediated RSA cryptography specification for additive private key splitting (mRSAA). *IETF Internet Draft, draft-kutylowski-mrsa-algorithm-03.* http://tools.ietf.org/html/draft-kutylowski-mrsa-algorithm-03.

[33] OASIS CMIS v1.1. (2013, May 23). Content Management Interoperability Services (CMIS). http://docs.oasis-open.org/cmis/CMIS/v1.1/CMIS-v1.1.html.

[34] OASIS DSS v1.0. (2010, November 12). *Profile for Comprehensive Multi-Signature Verification Reports Version 1.0.* Retrieved from http://docs.oasis-open.org/dss-x/profiles/verificationreport/oasis-dssx-1.0-profiles-vr-cs01.pdf

[35] OASIS-DSS. (2007, April 11). *Digital Signature Service Core Protocols, Elements, and Bindings Version 1.0.* Retrieved from OASIS Standard: http://docs.oasis-open.org/dss/v1.0/oasis-dss-core-spec-v1.0-os.html

[36] RFC 6238. (2011, July). Jerman Blazic, A.; Saljic, A.; Gondrom, T. *Extensible Markup Language Evidence Record Syntax (XMLERS).* https://tools.ietf.org/html/rfc6283.

[37] SD-DSS. (2011, August 09). *Digital Signature Service |Joinup.* Retrieved from https://joinup.ec.europa.eu/asset/sd-dss/description

[38] STORK 2.0. (2014). Retrieved from https://www.eid-stork2.eu/

[39] STORK. (2012). Retrieved from https://www.eid-stork.eu/

15

LEPS – Leveraging eID in the Private Sector

Jose Crespo Martín[1], Nuria Ituarte Aranda[1], Raquel Cortés Carreras[1],
Aljosa Pasic[1], Juan Carlos Pérez Baún[1], Katerina Ksystra[2],
Nikos Triantafyllou[2], Harris Papadakis[3],
Elena Torroglosa[4] and Jordi Ortiz[4]

[1]Atos Research and Innovation (ARI), Atos, Spain
[2]University of the Aegean, i4m Lab (Information Management Lab), Greece
[3]University of the Aegean, i4m Lab (Information Management Lab)
and Hellenic Mediterranean University, Greece
[4]Department of Information and Communications Engineering,
Faculty of Computer Science, University of Murcia, Murcia, Spain
E-mail: jose.crespomartin.external@atos.net; nuria.ituarte@atos.net;
raquel.cortes@atos.net; aljosa.pasic@atos.net; juan.perezb@atos.net;
katerinaksystra@gmail.com; triantafyllou.ni@gmail.com;
adanar@atlantis-group.gr; emtg@um.es; jordi.ortiz@um.es;

Although the government issued electronic identities (eID) in Europe
appeared more than 20 years ago, their adoption so far has been very low. This
is even more the case in cross-border settings, where private service providers
(SP) from one EU Member State needs trusted eID services from identity
provider located in another state. LEPS project aims to validate and facilitate
the connectivity options to recently established eIDAS ecosystem, which
provides this trusted environment with legal, organisational and technical
guarantees already in place. Strategies have been devised to reduce SP implementation costs for this connectivity to eIDAS technical infrastructure. Based
on the strategy, architectural options and implementation details have been
worked out. Finally, actual integration and validation have been done in two
countries: Spain and Greece. In parallel, market analysis and further options
are considered both for LEPS project results and for e-IDAS compliant eID
services.

15.1 Introduction

With the eIDAS regulation [1], the EU has put in place a legal and technical framework that obliges EU Member States to mutually recognize each other's notified eID schemes for cross-border access to online public-sector services, creating at the same time unprecedented opportunities for the private online service providers. The concept of notified eID limits the scope of electronic identity to the electronic identification means issued under the electronic identification scheme operated by one of the EU Member State (MS), under a mandate from the MS; or, in some cases, independently of the MS, but recognised and notified by that MS. To ensure this mutual recognition of notified e-ID, so called eIDAS infrastructure or eIDAS network has been established with an eIDAS node in each MS that serves as a connectivity proxy towards notified identity schemes. For a service provider that wants to connect to eIDAS network and to use cross-border eID services through it, this resolves only a part of the overall connectivity challenges. The connection from a service provider to its own MS eIDAS node, still has to be implemented by themselves and costs could be a considerable barrier.

This is where LEPS (Leveraging eID in the Private Sector) projects comes into picture. It is a European project financed by the EU through the Connecting Europe Facility (CEF) Digital programme [2], with a duration of 15 months, under grant agreement No. INEA/CEF/ICT/A2016/1271348. The CEF programme, with Digital Service Infrastructure (DSI) building blocks such as eID [3], aims at boosting the growth of the EU Digital Single Market (DSM). While public service providers are already under obligation to recognize notified eID services from another MS, private sector online service providers are especially targeted in CEF projects, and in LEPS in particular, in order to connect them to eIDAS network and offer eIDAS compliant eID services to the European citizens.

The LEPS consortium is formed by 8 partners from Spain and Greece. The project is coordinated by Atos Spain, that also performs the integration with the Spanish eIDAS node and supporting the Spanish partners. The University of Aegean in Greece performing the integration with the Greek eIDAS node and is supporting the Greek partners.

Three end-users participate in the projects in order to validate the use of the pan-European eIDAS infrastructure:

- Two postal services companies in Spain and Greece (Sociedad Estatal de Correos y Telégrafos and Hellenic Post respectively) integrating existing online services

- A digital financial services provider from Greece, Athens Exchange Group (ATHEX), aiming to offer remote electronic signature services to EU customers, compliant with eIDAS regulation.

Other partners include the Universidad de Murcia that creates the mobile application for using NFC eID cards, the Hellenic Ministry of Administrative Reconstruction in charge of the eIDAS node, and the National Technical University of Athens supporting the Greek partners.

Challenges in LEPS project cannot be understood outside of context of market adoption of "eIDAS eID services". However, set of challenges related to service provider (SP) connectivity to eIDAS is the main scope of the project. The focus is on the SP side of the eID market, more specifically sub-group of private sector online service providers. The approach taken in LEPS is to explore different options related to integration through so called eIDAS adapters in order to reduce burden for service providers and to reduce overall costs. eIDAS adapters is a sort of generic name given to reusable components, such as supporting tools, libraries or application programming interfaces.

The second group of challenges is around end user adoption, which indirectly affects service providers as well. Many service providers wait for the moment when citizens will activate and start to use massively their eID. This is also explored in LEPS project through design and development of mobile interface for the use of Spanish eID supporting NFC (known as DNI 3.0). The uptake of mobile ID solutions in many countries, notably in Austria, Belgium and Estonia, is growing faster than expected, so the introduction of LEPS mobile ID solution for Spanish DNI 3.0 can be considered as "right on time" action.

To summarise, challenges LEPS project faces are both related to the adoption of eIDAS eID services in general, as well as specific and related to the SP-to-eIDAS connectivity.

Finally, we can say that LEPS is fully aligned with the overall aim of CEF programme [4] to bring down the barriers that are holding back the growth of the EU Digital Single Market the development of which could contribute additional EUR 415 billion per year to EU economy.

15.2 Solution Design

The eIDAS network has been built by the European Commission (EC) and EU Member States based on previous development made in European projects such as STORK 1.0 [5] and STORK 2.0 [6]. The work developed by the eSENS project [7], and the collaboration with Connecting

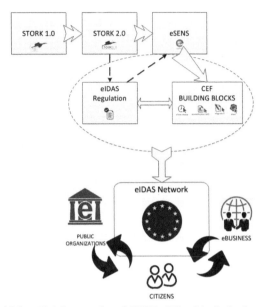

Figure 15.1 eIDAS network and CEF building blocks background [9].

Europe Facility (CEF) Digital [8] led to the generation of so called DSI building blocks "providing a European digital ecosystem for cross-border interoperability and interconnection of citizens and services between European countries" [9]. Figure 15.1 shows eIDAS network and CEF building blocks evolution.

While many public service providers, especially at the central government level, have already been connected to eIDAS network, although mainly in pilot projects and pre-production environments, the interest of private service providers to connect to eIDAS network and use eID services has been so far very limited. One of the main challenges, it has been mentioned, is related to the uncertainty about architectural options, costs and overall stability and security of the service provision.

With the aim to integrate the selected private SP services with the eIDAS network, two different approaches, one for Spain and one for Greece, were designed and implemented.

For the **Spanish services** scenario, the **eIDAS Adapter** [9] is the API implemented by ATOS which allows Correos Services (through MyIdentity service) to communicate with the Spanish eIDAS node. The eIDAS Adapter is based on a Java integration package provided by the Spanish Ministry for integrating e-services from the private sector with the Spanish eIDAS node.

This integration package, in its turn, uses the integration package delivered by the EC [10]. The Spanish eIDAS adapter provides a SP interface to the Correos' services, and an eIDAS interface for connecting to the eIDAS infrastructure, through Spanish eIDAS node, as depicted in Figure 15.2.

The eIDAS adapter is able to integrate Correos services with the eIDAS network, allowing a Greek citizen accessing to Correos e-services using a Greek eID, as is explained in Validation section.

For the **Greek services** scenario, the University of the Aegean has proposed a similar approach, as can be seen in Figure 15.3.

The integration of Greek services with the eIDAs network is made through the called LEPS eIDAS API Connector [11]. This API Connector re-uses the basic functionalities of eIDAS Demo SP package provided by CEF [10], and is provided in three different flavours, which can be used in different scenarios.

1. **eIDAS SP SAML Tools Library**. Used in the case of Java-based SP (developed from scratch) in which there's no need for one certificate for many services within SP and in which there is no need for pre-built UIs. This was used to avoid extra development time for creating and processing SAML messages.

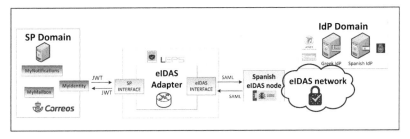

Figure 15.2   eIDAS adapter architecture general overview [9].

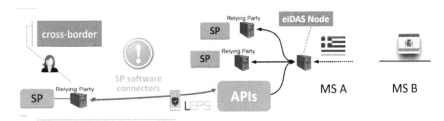

Figure 15.3   SP integration with eIDAS node using greek connector(s) [11].

2. **eIDAS WebApp 2.0**. This solution is for Java or non-Java-based SP scenarios, in which there is no need for one certificate for many services within SP but with need for built-in UIs. This allows to avoid development time for processing SAML messages, completely handles an eIDAS-based authentication flow (including UIs). Is SP infrastructure independent and operates over a simple REST API. This solution increases the security (JWT based security) (Figure 15.4).

3. **eIDAS ISS 2.0**. This solution is for Java or non-Java-based SP (developed from scratch) in which it is used one certificate for many services within SP and comes with or without SP e-Forms/thin WebApp. It is used to avoid development time for processing SAML messages, supports the interconnection of many SP services in the same domain (each service is managed via a thin WebApp). It sends SAML 2.0 request to eIDAS Node, translates response from SAML 2.0 to JSON and other common enterprise standards (and forward it to the relevant SP service). It's for multiple services with the same SPs sharing one certificate (Figure 15.5).

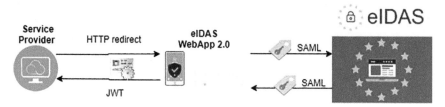

Figure 15.4   eIDAS WebApp 2.0 [12].

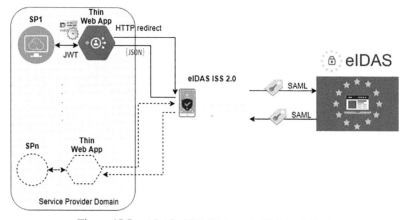

Figure 15.5   eIDAS ISS 2.0 (plus this WebApp) [12].

These connectors provided APIs facilitates the integration of ATHEX and Hellenic Post (ELTA) services with the eIDAS network, allowing a Spanish citizen accessing to Greek e-services using the Spanish eID, as indicated in Validation section.

15.2.1 LEPS Mobile App

The use of smartphones and tablets for interacting with public administrations and private companies has currently become an increasing common practice, therefore it is necessary to offer mobile solutions that integrate mobile eIDAS authentication in the SP service ecosystem.

An efficient solution for **mobile devices** with a successful integration of mobile eIDAS authentication is offered. Concretely, mobile app provides mobile support for Greek services (ATHEX and ELTA) to enable authentication of Spanish citizens, through eIDAS infrastructure using the Spanish DNIe 3.0 (electronic Spanish Identity National Document), which supporting NFC technology [13].

The mobile application developed by Universidad de Murcia can work with any SP offering eIDAS authentication for Spanish users [14]. Additionally, the implementation can be easily extended to other EU Member States by adding other authentication methods beyond the Spanish DNIe. Also, the requirements for SPs to integrate the mobile application are practically minimal and limited to the global requirements of operating in a mobile environment, i.e. providing responsive interfaces and use standard components such as HTML and JavaScript 7.

15.3 Implementation

Aiming to cover all the functionalities and requirements needed by the SPs and the eIDAS network, the developed **Spanish eIDAS adapter** comprises the following modules depicted in Figure 15.6:

- **SP interface**: Establishes interaction with the integrated services. Contains a single endpoint which receives the authentication request from the SP;
- **eIDAS interface**: Connects to the country eIDAS node, Comprises two endpoints:
  - Metadata endpoint: Provides SP metadata;
  - ReturnPage endpoint: Receives the SAML response from the country eIDAS node.

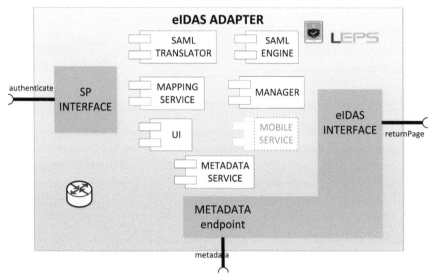

Figure 15.6   Spanish eIDAS adapter modules [10].

- **UI module**: Interacts with the end user;
- **Manager service**: Orchestrate the authentication process inside the eIDAS Adapter;
- **Translator service**: Translates in both ways from the SP to eIDAS node:
    - The authentication request (JWT) from the SP to a SAML request;
    - The SAML response from eIDAS node to an authentication response (JWT) to SP;
- **Mapping service**: Maps the SP attribute names to SAML eIDAS attribute names, doing the semantic translation;
- **SAML Engine**: Manages the SAML request and response, encrypting/decrypting and signing;
- **Metadata service**: Creates SP metadata;
- **Mobile service:** Optional component able to detect the device where the authentication process is performed.

The most relevant technologies, standards and protocols used during the implementation include among others: Java 8.0 as implementing language, JWT (industry standard RFC 751) to transmit the user data between the SP and the adapter in a secure way, SAML 2.0 for transmitting user authentication between the eIDAS infrastructure and the adapter. For the deployment process Apache Tomcat as web application server was used and deployed as a Docker container.

During the implementation, deployment and test of the Spanish eIDAS adapter some challenges arose. The actions performed for overtaking these challenges could help MS for taking decisions when facing the implementation of new adapters for integrating private online services to eIDAS infrastructure.

The plan for designing and implementing this adapter was reuse the integration package the Spanish Ministry provided for the private SP integration. This approach would help to reduce the use of resources guarantying the connection to eIDAS node. Thus, only some minor effort for integrating SP service would be needed. Despite this advantage the use of legacy code and the used technologies could restrict the use of cutting-edge or more familiar. In the particular case of the Spanish eIDAS adapter implementation, mixing technologies such as Struts 2 and Spring took more time than expected. It was necessary to carry out some changes and the developer team had to acquire some knowledge on Struts 2 framework. The use of generic eIDAS libraries beside well know technologies by the development team, is recommended. Apart from this, the relevant integration information and the technical support the organization in charge of the country eIDAS node can provide, is very useful.

As a summary the main features of the Spanish eIDAS adapter [10] are:

- Modular design;
- Reusable;
- JWT based security for transmitting user information;
- Able to create SAML Requests and process SAML Reponses;
- Translate SAML 2.0 to JSON and vice versa;
- SP client programming language independent;
- Docker based deployment.
- SP infrastructure independent (can be deployed on SP infrastructure or on a third party);
- Able to connect with different SP services in the same or from different domains.

Regarding the **Greek API connectors** [11]:

The **eIDAS SP SAML Tools library** can be used to simplify SP-eIDAS node communication development, on the SP side. It is offered in the form of a Java library, which can be easily integrated into the development of any Java-based SP. The library itself is based on the CEF provided SP implementation (demo SP). This library provides methods that a Java-based SP implementation can call to create SAML Requests (format, encode, encrypt), parse SAML Responses (decrypt, decode, parse) and create the SP metadata xml, as required by the eIDAS specifications [11].

The **eIDAS WebApp 2.0** uses the previous eIDAS SP SAML Tools library, providing a UI, a simple REST API and the business logic for handling the eIDAS authentication flow. The WebApp is offered as a Docker image for deployment purposes and need to be deployed on the same domain than the SP [11].

The **eIDAS ISS 2.0** simplifies the connection of any further SP enabling SPs to connect to the eIDAS node without using SAML 2.0 protocol. Allowing that one ISS 2.0 installation can support multiple services within the same SPs. Provides communication endpoint based on JSON. The ISS 2.0 app is provided as a war artefact to be deployed on Apache Tomcat 7+ [11].

15.4 Validation

With the aim to demonstrate and validate the SP integration with the eIDAS infrastructure through the country eIDAS nodes, the following selected services have been customized in order to proceed with the integration.

The selected services customized on the Spanish side were provided by Correos [15]:

- "My Identity" provides secured digital identities to citizens, businesses and governments;
- "My Mailbox" is a digital mailbox and storage that enables you to create a nexus of secure document-based communication;
- "My Notifications" provides a digital service that aims to centralize and manage governmental notifications.

The services provided on the Greek side were provided by ATHEX and ELTA [16]:

- "Athex Identity service" provides an eIDAS compliant identity provider service;
- "Athex Sign" is a service that provides a secure way to sign on the go. Anytime and anywhere;
- "Athex AXIAWeb" allows any European Union citizen-investor to register and login via eIDAS.
- "ELTA e-shop" offers functionalities such as letter mail services or prepaid envelopes;

- "ELTA eDelivery Hybrid Service" provides document management functionalities through the use of digital signatures, standardization flow and other tools;
- "Parcel Delivery Voucher" allows customers print online the accompanying vouchers for parcels send to their customers.
- "Online Zip Codes" allows corporate customers to obtain the current version of Zip codes of Greece.

After the customization and integration, the IT infrastructure of these services were connected to the appropriate country eIDAS node, allowing the services to use the eIDAS network for user authentication with eID issued by EU Member States. Additionally, is demonstrated the usability of eIDAS specifications and the Spanish and Greek eIDAS nodes in the private sector.

For testing the integrated services during the project, the following steps have been performed [17]:

1. Preparation of the pre-production tests necessary for the SP services integration verification in pre-production environment;
2. Execution of the pre-production automated tests against the Spanish (for Correos) and Greek (for ATHEX and ELTA) eIDAS node using test credentials on a pre-production environment. Finally, feedback of this step is generated for the next steps for the production testing;
3. Preparation of the production tests considering the feedback from previous steps;
4. Execution of the production manual tests against the Spanish eIDAS node in pre-production environment (for pre-production Correos services) due to Spanish Ministry restrictions, and against (for production ATHEX and ELTA services) production eIDAS Node. In both cases real credentials were used.

The automated testing on pre-production environment where performed using an automated testing tool eCATS (eIDAS Connectivity Automated Testing Suite) based on Selenium Selenium portable software-testing framework for web applications. eCATS tool has been customized for each integrated service as depicted in Figure 15.7.

Apart from the connectivity tested between the Spanish and the Greek eIDAS nodes, additional interoperability tests have been performed from the Spanish eIDAS node to Iceland, The Netherlands and Italy, and between Greece and the Czech Republic. For this purpose, test credentials provided by different public organizations in change of the eIDAS node management in their countries were used.

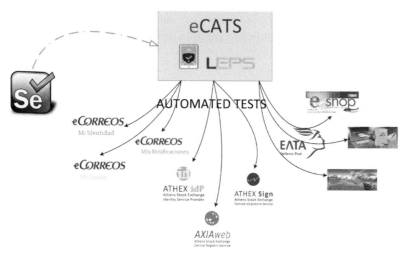

Figure 15.7   LEPS services and automated eCATS tool [13].

15.5 Related Work

LEPS project is linked to a set of projects where the LEPS partners participated with the aim to increase the use of the eID between the EU citizens for accessing online services across EU and reinforce the Digital Single Market in Europe. Among these is worth to mention the following projects:

- **STORK** [5] (CIP program, 2008–2011), providing the first European eID Interoperability Platform allowing citizens to access digital services across borders, using their national eID;
- **STORK 2.0** [6] (CIP program, 2012–2015), as continuation of STORK was intended to boost the acceptance of eID in EU for electronic authentication for both physical and legal person, and place the basis for the creation of an interoperable and stable cross-border infrastructure (eIDAS network) for public and private online services and attribute providers;
- **FutureID** [18] (FP7-ICT program, 2012–2015), "created a comprehensive, flexible, privacy-aware and ubiquitously usable identity management infrastructure for Europe, integrating existing eID technology and trust infrastructures, emerging federated identity management services and modern credential technologies to provide a user-centric system for the trustworthy and accountable management of identity claims" [18];

- **FIDES** [19] (EIT, 2015–2016), built a secure federated and interoperable identity management platform (mobile/desktop). An identity broker was implemented, where STORK infrastructure was provided;
- **STRATEGIC** [20] (CIP program, 2014–2017), provided more effective public cloud services, where STORK network was integrated;

Additionally, there are projects related to eID management in different sectors, such as the academic domain, or the public sector, where the LEPS partners are also participating, and are currently under development.

- **ESMO** project aims to integrate education sector Service Providers to the eIDAS network, contributing to increase eIDAS eID uptake and use in the European Higher Education Area (EHEA) [21]. Outcomes from LEPS will be used;
- **TOOP** [22] project main objective is "to explore and demonstrate the once-only principle across borders, focusing on data from businesses". TOOP demo architecture implementation incorporates LEPS APIs (WebApp 2.0). Task: Identify users accessing the services of a TOOP Data Consumer via eID_EU.
- **FIWARE** project (FP7-ICT) [23]. Since Atos is co-founder of FIWARE Foundation it is supporting publication of connectivity to CEF eID building block as generic enabler. LEPS adaptor for Spanish DNIe will be part of know-how exchange with Polytechnic University of Madrid, partner responsible for generic enabler.

Finally, LEPS project stablished links with other eID projects under the umbrella of the LEPS Industry Monitoring Group (IMG), such as "Opening a bank account with an EU digital identity" CEF Telecom eID project [24] and "The eIDAS 2018 Municipalities Project CEF Telecom eID project" [25]. Contacts have also been done with other eID initiatives such as Future Trust, ARIES and Credential projects, as well as industrial initiatives EEMA, ECSO, TDL, OIX, Kantara and OASIS.

Besides direct integration of external eID services through the identity provider available APIs, e-service providers have also an option to use broker or aggregator of different identity providers that might offer additional eID services or functionalities. In this category we can mentioned related work on so-called Identity clouds or CIAM (customer identity and access management) solutions. In Germany SkIDentity Service [26] is a kind of broker for service providers that can use popular social logins such as LinkedIn and Facebook Login, as well as eIDAS eID services from a number of countries. In Netherlands, similar broker role to municipal e-services is provided by Connectis with support from CEF project. In Spain Safelayer has offering named TrustedX eIDAS Platform that is "orchestrating" digital identities for authentication, electronic signature, single sign-on (SSO) and Two-factor Authentication (2FA) for Web environments.

15.6 Market Analysis

LEPS Market analysis had to take into account specific pre-existing context of provide e-service provider, such as:

1. Organisations that need or want to make migration from the existing identity and access management (IAM) solution. This could apply to organisations that have scaled out their internal or tailor made IAM solutions, or organisations that already use partially external or third-party e-identification or authentication services, but are looking for the services with a higher level of assurance (LoA)
2. Organisations that use low assurance third party eID services, such as social login, and want to elevate overall level of security and decrease identity theft and fraud by integration of eIDAS eID services, either to replace or to enhance exiting external eID services
3. Organisations that are already acting or could be acting as eID brokers
4. Organisations that want to open new service delivery channels through mobile phone and are interested in mobile ID solutions that work across borders

The first group is composed of organisations that made important investments in their internally operated IAM solutions. These solutions, however, are originally not meant to handle the requirements for large scale cross-border e-service use cases, although the functional building blocks and protocols might be the same. They usually have internal know-how and capacity to implement eIDAS connectivity and the main driver for adoption of eIDAS eID services could be regulatory compliance, such as "know your customer" (KYC) requirement in anti-money laundry (AML) directive.

In addition, given that the main value proposition of LEPS approach, right from the start, was based on cost effectiveness and, in a lesser amount, also on cost efficiency, one of the main adoption targets are small and medium enterprises (SME) operating in cross-border context, planning migration or extension of their current third-party eID services. Unlike the first group of LEPS adopters, these organisations are unlikely to have know-how, resources and capacity to implement eIDAS connectivity. The main proposition from LEPS in this regard is saving cost and time for e-service provider organisations in regard to activities such as familiarization with SAML communication (protocol understanding and implementation), implementation of the required web interface (UI) for user interaction with the eIDAS-enabled services, formulation and proper preparation of an eIDAS SAML Authentication Request, processing of an eIDAS Node SAML Authentication Response and provision of the appropriate authentication process end events for success or failure.

The fact that many organisations do not have resources for eID service implementation and operation internally was already exploited by social networks and other online eID service providers that offer their "identity APIs". This is an easy way to integrate highly scalable, yet low assurance, eID services. In some SP segments, such as e-commerce, there is a huge dominance of Facebook and Google eID services (with 70% and 15% market share respectively), while in the other segments, so called customer IAM (or CIAM) appeared as an emerging alternative to integrate API gateways to different online eID service providers.

This new generation of CIAM solutions, complemented by a variety of eID broker solutions, is the third potential target for LEPS adoption. Integrating external identities can be linked to onboarding, such as in the case of Correos Myidentity service, or can help in trust elevation and/or migration from social e-IDs with low LoA to eIDAS eID with high LoA. With scalability, there are other requirements that might depend on a specific e-service provider, such as for example integration with customer relationship management or handling a single customer with many identities.

From all trends that have been analysed in market analysis, the one that is most promising to impact LEPS results uptake is mobile identification and authentication, which targets user experience and usability. Given that the subset of LEPS results also contains interface for mobile eID (although only Spanish DNI 3.0), the organisation that have this specific need, targeting Spanish citizens that use mobile e-service from other members state SP, are considered as the fourth group of adopters.

For all of these users, LEPS brings benefits of cost saving, while eIDAS eID services represent well known benefits:

- Improved quality of the service offered to the customer;
- The introduction of a process of identity check and recognition through eID reduces frauds;
- Reduction in operational, legal and reputational risk as trusted identities and authentication is provided by national public MS infrastructures;

- Time savings, reduction in terms of administrative overhead and costs;
- Increasing potential customer base.

These theoretical assumptions have been partially validated in the case of LEPS service providers. In ELTA case, for example, possible users are Greek nationals living abroad and using some eID different than Greek. According to the General Secretariat for Greeks Abroad more than 5M citizens of Greek nationality live outside the Greek border, scattered in 140 countries of the world. The greater concentration has been noted in the US (3M), Europe (1M), Australia (0, 7M), Canada (0, 35 M), Asia – Africa (0, 1M) and Central and South America (0, 06M). In this view as regards cross-European e-delivery, the primary target for this service has customer base of 1M with initial penetration rate set to 1% (10.000 users). ELTA focus on existing and new customers was distinguished with 2 supplementary strategies: Revenue Growth (existing) and Market Share (new) Strategies.

As we can see from Table 15.1 (with data collected from the actual pilots), the cost of implementation of eIDAS connectivity depends of a selected architectural and software options. Reuse of LEPS results significantly reduces this cost, both for fixed one-time expenditures and for operational costs.

Two strategies envisaged by ELTA aimed at benefit of 100.000 within 2 years. With the figures from Table 15.1 it is clear that this breakeven point can be reached only with the reuse of LEPS components (with the accumulated cost of 87.624 euros for 24 months), while building eIDAS connectivity from the scratch would reach this point only in the year 3 (the accumulated cost at the end of 24-months period would be 111.235 euros for this option).

Table 15.1   Cost of three eIDAS connectivity options

Integration Scenario	Fixed Cost (in EUR)	Operational Cost (in EUR per year)
Build from scratch (Scenario 1)	41,739.36	34,748.44
Build by using CEF Demo SP (Scenario 2)	33,792.96	32,627.80
LEPS eIDAS API Connectors (Scenario 3)	25,734.46	30,945.28

15.7 Conclusion

The challenge of adoption of eIDAS eID services can be divided into challenges related to service provider connectivity with eIDAS network and challenges related to citizen/business use of notified eID means, including NFC enabled eID cards in case of Spanish citizens. LEPS projects tried to reduce gaps for both types of challenges. The solution for service provider connectivity to eIDAS nodes can be considered as easily replicable across EU. It is focused on service provider cost saving, when it comes to investments in eIDAS connectivity. The other LEPS solution, focused on the Spanish eID card use through mobile phone interface is targeting usability as the main value proposition.

The analysis of architectural option provided by LEPS project demonstrated that there are different approaches to integrate online services with the eIDAS infrastructure through the connection with the country eIDAS node. Implementation of API's in two countries and pilot trials with real services and users, resulted not only in technical verification of selected approaches, but also in validation from cost-benefit and usability perspectives.

The final outcomes generate benefits mainly to the SPs such as reduction of time and effort for integrating their online services with the pan-European eIDAS network. In its turn, the use of eIDAS eID services facilitates the cross-border provision of e-services and elevates level of trust by end users. In addition, LEPS interface for mobile access to Spanish DNI3.0 eID card services improves the user experience. Finally, the results of the project will benefit larger community of eIDAS developers and other stakeholders since results and guidelines generated during the project will help in taking decisions on how to approach and manage the challenges related to service provider connectivity to the country eIDAS node.

Acknowledgements

This project has been funded by the European Union's Connecting Europe Facility under grant agreement No. INEA/CEF/ICT/A2016/1271348.

References

[1] EUR-Lex, Access to European Union law, Regulation (EU) No. 910/ 2014 of the European Parliament and of the Council of 23 July 2014 on electronic identification and trust services for electronic transactions

in the internal market and repealing Directive 1999/93/EC, https://eur-lex.europa.eu/legal-content/EN/TXT/?uri=uriserv%3AOJ.L_.2014.257.01.0073.01.ENG. Retrieved date 30 November 2018.

[2] European Commission, CEF Digital, https://ec.europa.eu/cefdigital/wiki/ display/CEFDIGITAL/CEF+Digital+Home. Retrieved date 27 November 2018.

[3] European Commission, About CEF building blocks, https://ec.europa.eu/ cefdigital/wiki/display/CEFDIGITAL/About+CEF+building+blocks. Retrieved date 28 November 2018.

[4] European Commission, Digital Single Market, https://ec.europa.eu/digi tal-single-market/e-identification. Retrieved date 26 November 2018.

[5] European Commission, STORK: Take your e-identity with you, every-where in the EU, https://ec.europa.eu/digital-single-market/en/content/ stork-take-your-e-identity-you-everywhere-eu/. Retrieved date 30 November 2018.

[6] European Commission, Digital Single Market, STORK 2.0, https://ec.eu ropa.eu/digital-single-market/en/news/end-stork-20-major-achievements-making-access-mobility-eu-smarter. Retrieved date 30 November 2018.

[7] eSENS, Moving Services Forward, https://www.esens.eu/. Retrieved date 30 November 2018.

[8] European Commission, e-SENS and Connecting Europe Facility: how do they work together? https://ec.europa.eu/digital-single-market/en/ news/e-sens-and-connecting-europe-facility-how-do-they-work-together. Retrieved date 30 November 2018.

[9] D3.3 Operational and Technical Documentation of SP integration. Lead author: Juan Carlos Pérez Baún, Deliverable of the LEPS project, 2018. https://leps-project.eu/node/345. Retrieved date 30 November 2018.

[10] European Commission, eIDAS-Node integration package, https://ec.euro pa.eu/cefdigital/wiki/display/CEFDIGITAL/eIDAS+Node+integration+ package. Retrieved date 18 November 2018.

[11] D4.2 eIDAS Interconnection supporting Service, Lead authors: Petros Kavassalis, Katerina Ksystra, Nikolaos Triantafyllou, Harris Papadakis, Deliverable of LEPS project 2018, http://www.leps-project.eu/node/347, Retrieved date 30 November 2018.

[12] D4.3 Operational and Technical Documentation of SP (ATHEX, Hellenic Post) integration (production), Lead authors: Petros Kavassalis, Katerina Ksystra, Nikolaos Triantafyllou, Maria Lekakou, Deliverable of LEPS project 2018, https://leps-project.eu/node/348. Retrieved date 30 November 2018.

[13] DNI y Pasaporte, Cuerpo Nacional de Policia, DNI electrónico, Descripción DNI 3.0, https://www.dnielectronico.es/PortalDNIe/PRF1_Cons02.action?pag=REF_038&id_menu=1. Retrieved date 30 November 2018.

[14] D3.1 – Mobile ID App and its integration results with the Industrial Partners. Lead author: Elena Torroglosa, Deliverable of the LEPS project, 2018.https://leps-project.eu/node/343. Retrieved date 30 November 2018.

[15] D3.2 Operational and Technical Documentation of Correos services customization. Lead author: Juan Carlos Pérez Baún http://www.leps-project.eu/node/344. Retrieved date 30 November 2018.

[16] D4.1 Operational and Technical Documentation of SP (ELTA, ATHEX) customization. Lead author: Petros Kavassalis, Katerina Ksystra. http://www.leps-project.eu/node/346. Retrieved date 30 November 2018.

[17] D6.1 Production Testing Report. Lead authors: Petros Kavassalis, Katerina Ksystra, Manolis Sofianopoulos. http://www.leps-project.eu/node/354. Retrieved date 30 November 2018.

[18] FutureID, Shaping the future of electronic identity, http://www.futureid.eu/ Retrieved date 30 November 2018.

[19] FIDES, Federated Identity Management System, EIT Digital, https://www.eitdigital.eu/fileadmin/files/2015/actionlines/pst/activities2015/PST_Activity_flyer_FIDES.pdf. Retrieved date 30 November 2018.

[20] STRATEGIC, Service Distribution Network and Tools for Interoperable Programmable, and Unified Public Cloud Services, https://ec.europa.eu/digital-single-market/sites/digital-agenda/files/logo_strategic_v3_final.jpg. Retrieved date 10 December 2018.

[21] ESMO, e-IDAS-enabled Student Mobility, http://www.esmo-project.eu/. Retrieved date 30 December 2018.

[22] TOOP, The Once-Only Principle Project, http://www.toop.eu/. Retrieved date 10 December 2018.

[23] FIWARE, Future Internet Core Platform, https://www.fiware.org/. Retrieved date 10 December 2018.

[24] CEF: Opening a Bank Account Across Borders with an EU National Digital Identity,https://oixuk.org/opening-a-bank-account-cross-border-id-authentication/ Retrieved date 10 December 2018.

[25] CEF: The eIDAS 2018 Municipalities Project, https://ec.europa.eu/cef digital/wiki/display/CEFDIGITAL/2017/07/11/eIDAS+2018+Municipalities+Project. Retrieved date 10 December 2018.

[26] Secure identities for the web: Skidentity web site: skidentity.com

Index

About the Editors

Dr. Jorge Bernal Bernabe received the MSc, Master and PhD in Computer Science from the University of Murcia. He was accredited as Associate Professor by Spanish ANECA in 2016 and granted with the "Best PhD Thesis Award" from the School of Computer Science of the University of Murcia 2015. Currently, he is a Postdoctoral researcher in the Department of Information and Communications Engineering of the University of Murcia, partially supported by INCIBE (Spanish National Cybersecurity Institute). Jorge Bernal has been visiting researcher in the Cloud and Security Lab of Hewlett-Packard Laboratories (Bristol UK) and in the University of the West of Scotland. Author of several book chapters, more than 20 papers in indexed impact journals and more than 20 papers in international conferences. He has been involved in the scientific committee of numerous conferences and served as a reviewer for multiple journals. During the last years, he has been working in several European research projects related to security and privacy, such as POSITIF, DESEREC, Semiramis, Inter-Trust, SocIoTal, ARIES, ANASTACIA, OLYMPUS, CyberSec4Europe. His scientific activity is mainly devoted to the security, trust and privacy management in distributed systems and IoT.

Dr. Antonio Skarmeta received the M.S. degree in Computer Science from the University of Granada and B.S. (Hons.) and the Ph.D. degrees in Computer Science from the University of Murcia Spain. Since 2009 he is Full Professor at the same department and University. Antonio F. Skarmeta has worked on numerous research projects in the national and international area in the networking, security and IoT area, such as Euro6IX, ENABLE, DESEREC, Inter-Trust, DAIDALOS, SWIFT, SEMIRAMIS, SMARTIE, SOCIOTAL, IoT6, ARIES, ANASTACIA, OLYMPUS, CyberSec4Europe. His main interest is in the integration of security services, identity, IoT and Smart Cities. He heads the research group ANTS since its creation on 1995. Currently, he is also advisor to the vice-rector of Research of the University of Murcia for International projects and head of the International Research Project Office. Since 2014 he is Spanish National Representative for the MSCA within H2020. He has published over 200 international papers and being member of several program committees. He has also participated in several standardization for a like IETF, ISO and ETSI.